CHAINSAW LOVE

CHAINSAW LOVE

FIELD NOTES ON THE WORLD'S MOST DYNAMIC POWER TOOL

JAMES CARD

WISCONSIN HISTORICAL SOCIETY PRESS

Published by the Wisconsin Historical Society Press
Publishers since 1855

The Wisconsin Historical Society helps people connect to the past by collecting, preserving, and sharing stories. Founded in 1846, the Society is one of the nation's finest historical institutions.
Join the Wisconsin Historical Society: wisconsinhistory.org/membership

Publication of this book was made possible in part by a gift from Midwest STIHL.

Front and back cover illustrations by Andrew Brozyna.

Unless otherwise noted, all photographs in the book were provided courtesy of the author.

Printed in Canada
Typesetting by John Ferguson

29 28 27 26 25   1 2 3 4 5

Library of Congress Cataloging-in-Publication Data

Names: Card, James (Journalist) author
Title: Chainsaw love : field notes on the world's most dynamic power tool / James Card.
Description: Madison, WI : Wisconsin Historical Society Press, [2025] | Includes bibliographical references and index.
Identifiers: LCCN 2024061785 (print) | LCCN 2024061786 (e-book) | ISBN 9781976600494 paperback | ISBN 9781976600500 epub
Subjects: LCSH: Chain saws—Popular works | Sawing—Popular works | Wood-carving—Popular works | Logging—Machinery—Popular works
Classification: LCC TS850 .C33 2025 (print) | LCC TS850 (e-book) | DDC 621.9/34—dcundefined
LC record available at https://lccn.loc.gov/2024061785
LC e-book record available at https://lccn.loc.gov/2024061786

∞ The paper used in this publication meets the minimum requirements of the American National Standard for Information Sciences—Permanence of Paper for Printed Library Materials, ANSI Z39.48-1992.

Nature is not, as in the urbanist vision, the local well-manicured park flora or the evening weather forecast, but a brutal, uncaring world kept at bay with tractors, chainsaws, and sump pumps.

—VICTOR DAVIS HANSON

I go out with my chainsaw for a couple of hours, then write for a couple of hours.

—BERND HEINRICH

CONTENTS

Author's Note

When I use the word *saw* in casual conversation, I mean *chainsaw.* The same goes for this book: a saw is a chainsaw, and other types of saws like handsaws, polesaws, or crosscut saws are mentioned by their specific names. Among foresters, wildland firefighters, arborists, woodcarvers, and other cognoscenti of the bar and chain, when they speak of a saw, they speak of a chainsaw.

A few other notes on terminology: The act of cutting down a tree so it lands on the ground can be called felling or falling, depending on where you are from. "Falling a tree" is a colloquialism of loggers from America's West Coast. "Felling a tree" is used by about everyone else. In this book, I use *felling* as a verb and refer to the person performing the felling as a faller, which is the term used by the National Wildfire Coordinating Group and some of its partners, including the US Forest Service.

This is not an instructional book about using chainsaws, though many passages describe how a chainsaw is used. Rather, this book is a collection of insights, observations, and reflections about the world of chainsaws and profiles of people who use and appreciate them. Within the book are useful tips, recommendations, projects, and maintenance protocols meant to enrich and educate the reader and improve their experience using chainsaws. For learning how to use a chainsaw safely and properly, nothing beats training.

In some ways, chainsaws are like guns and motorcycles—many people believe they could pick up a Glock for the first time and hit a moving target like John Wick or hop on a superbike and take corners like the racers in the Isle of Mann TT, the most dangerous motorcycle race on earth. An observational familiarity with these machines, however, is not the same as the familiarity that comes with practice. The chainsaw is an amazing tool, but it is also quite dangerous. When blood is spilled, it is often from operators who did not respect the power of the machine and did not take the time to fully understand it.

I used to be one of those operators. I grew up cutting firewood with my father and developed a fascination with chainsaws. We probably broke every single chainsaw safety rule in existence. Luckily, neither of us was injured. To be fair, not many other people in the 1980s were safely using chainsaws either. When studying black-and-white photos and advertisements from the early days of chainsaw development, I noticed that nobody wore protective equipment—not even a pair of shop goggles. The development of built-in safety features, protective gear, and safe-handling concepts took a long time to evolve and become common practice among the general population.

It wasn't until 2019, when I received professional training as a line-clearance tree trimmer, that I realized my ignorance. As I worked my way through the Line Clearance Qualification Standard apprenticeship, I studied a dictionary-sized tome filled with chapters about chainsaws, arboriculture, tree climbing, rigging, and information about the US electrical grid. We practiced everything from CPR to rescuing injured climbers from trees to tying Distel hitches, but our world revolved around operating chainsaws. As I mastered each skill and learned each lesson, a foreman would confirm that I'd completed the corresponding section. I enjoyed this type of book learning combined with hands-on field experience. Most importantly, I carried those lessons home and applied them to the chainsaw work I performed around my homestead in the woods. I am much safer and more productive because of this training.

In Wisconsin, the Department of Natural Resources offers a chainsaw class for beginners, and the conservation departments of other states offer similar instruction. Universities, their extensions, and technical colleges sometimes have chainsaw classes for the general public. The US Forest Service has a National Crosscut and Chainsaw Program with four sawyer certification levels. The National Park Service has classes for volunteers. The Occupational Safety and Health Administration (OSHA) offers an online course, which isn't as good as hands-on learning, but it's an easy starting point. If you live near a major hiking trail, like the Ice Age Trail or the Pacific Crest Trail, nonprofit groups that maintain the trails frequently offer chainsaw safety classes for their volunteers. When trees topple over onto these trails, a small army of volunteer sawyers is needed to clear the paths. Nature centers sometimes offer classes, too.

Safety and Woods Worker (SAWW) training has four levels, and their trainers, who are scattered across the US, often host classes regionally. The

Forest Industry Safety & Training Alliance (FISTA) also offers four levels of training classes. There are also private groups that offer top-notch training, such as ACRT Arborist Training and Northeast Woodland Training, whose Game of Logging curriculum was created in the 1960s by Soren Eriksson, a Swedish logger.

I do not endorse any particular brand of chainsaw. Users of Stihl and Husqvarna saws are like the Ford versus Chevy loyalists who endlessly razz each other.

Husqvarna user: "I'm going to cut down that big tree over there."

Stihl user, holding out a hatchet: "Take this. It will be faster."

Both Stihl and Husqvarna make wonderful saws, as do many other companies. Several times in the book, I mention specific products and that is because I genuinely like them and find them useful.

1

Apocalypse Firewood

The Chainsaw as a Force Multiplier for Energy Procurement

Serious firewood stackers live by the concept of getting at least three years ahead. That means having a three-year inventory of firewood ready to burn. Year-one firewood should be seasoned and ready to burn at any given time of the year, while year-three wood may be green and will be rotated to burn two winters later. The reason for getting three years ahead is that next year you might have a bad back, a bad divorce, or a bad work schedule; you may not be able to spend much time cutting, splitting, and stacking firewood. Three years ahead means having a buffer for the bad times. It means saving wood for an unproductive year, as you would save money for a rainy day. As long as the wood is covered, it can last a very long time with little deterioration. Also, wood smoke is my aromatherapy, and a large and neatly stacked woodpile is pleasing to my eye.

Hitting the three-year goal is a significant personal achievement, not much different from setting a personal record for a deadlift, running a first marathon, or inventing a product. It is the result of slow, steady progress that builds on itself. Some wood stackers get five or ten years ahead. What constitutes a year's supply is relative, however, depending on how much wood a person burns.

To the owner of an outdoor furnace whose house is warmed entirely by wood heat, the door of that furnace likely resembles the gaping maw of a hungry fire-bellied dragon. A huge stash of wood is required to feed this beast. To the owner of a living room fireplace who burns wood on a casual basis, the fireplace may be like a household pet that needs some, but not much, wood to survive. A cord is the standard unit of a firewood stack, measuring four feet high, four feet wide, and eight feet long. A face cord is the same height and length, but the depth is shorter: a single stacked row. There isn't a standard length for firewood. Some people like small, football-sized chunks, and others prefer to maximize the length of their wood based on the dimensions of their firebox.

The wood-burning season is different for everyone. In my Wisconsin home, I often stoke the first fire sometime in October (trick-or-treating in the snow is not unusual here) and let the last fire die down in early or mid-May (opening day of Little League is sometimes played amid snowflakes in the breeze).

When I got two years ahead on my firewood stash, I was not just satisfied but also enormously relieved. I was tired of cutting, splitting, and stacking wood. I needed a break. But a couple of months later, I was back in the woods felling some barkless oaks that had died years ago. I believe that turning these old trees into firewood honors them. I toss the smaller limbs into a brush pile that will provide cover for rabbits, raccoons, possums, and skunks.

Cutting firewood is in the same philosophical school of thought-put-to-action as gardening, farming, hunting, and fishing. By growing vegetables, raising chickens, butchering deer, and filleting walleye, you put yourself back into the food chain as opposed to being a passive consumer of food grown in faraway places and purchased wrapped in plastic from a grocery store. It is an act of rebellion, a refusal of the Soylent Green that Big Ag foists upon us. Similarly, by cutting and burning firewood, you insert yourself back into the energy chain, no longer beholden to the utilities and the oil and gas conglomerates. The stored energy of wood is extracted locally. You reclaim your energy independence, and the warmth of your home is not held hostage by international syndicates. The official state motto of West Virginia is *Montani Semper Liberi*, which translates to Mountaineers Are Always Free. My personal motto is similar: *Lignatoris Semper Liberi*. Woodsmen Are Always Free.

When I perform manual labor, I often think about a particular question or subject as I'm working. It is much like the idea of a Japanese Zen koan—a paradox, story, riddle, or question used as a meditation tool. In my form of working meditation, however, instead of sitting on my ass in

A stack of firewood is a manifestation of self-reliance. It is a locally sourced, free range, organic, and sustainable fuel. And seasoning it requires only the sun, the wind, and time.

the lotus position, I am meditating *and* being productive while stacking firewood. Using this technique, I've had many aha moments and creative breakthroughs while doing some of the shittiest jobs. The more mindless the work, the better the insights.

A recurring question that often comes to mind while I'm making firewood: "What if I had to do all of this work without a chainsaw?" This gets me thinking about the pioneers and the settlers and the lumberjacks of legend.

This, in turn, leads me to wonder: what could prevent me from using my chainsaw? If it broke down, I'd fix it. But it could be out of gas. What would cause a lack of gas? A supply-chain breakdown or fuel shortages or peak oil could actually come true. If there was no electrical power, the pumps would not work, and no electrical power would mean that the grid somehow went down. I think back to my time doing line-clearance tree-trimming work and all of the close calls we had while felling trees near high-voltage power lines. The grid could go down for any number of reasons, and that would put modern society on the brink of collapse.

Such things have been on my mind since the start of the COVID-19 pandemic. In the middle of a crisis, many people look to go back to the land and to become more self-sufficient. When the world as we know it is turned upside down, lots of us feel a natural instinct to become stronger and more independent.

If the grid were to go down, most people with woodstoves or fireplaces would naturally think they'd be secure with a source of energy to heat their homes. However, few people understand the labor involved in putting up a stockpile of wood that can be burned twenty-four hours a day for months at a stretch. Firewood can be bought and sold, and there are woodcutters in every community with a few cords for sale. But in a grid-down situation, those cords would be in high demand and would come at a very steep price—assuming the mass of wood could even be picked up or delivered. In an apocalyptic event, cutting wood would be one of the last things you'd want to do, as it takes time and huge amounts of physical energy. Time would be better spent accumulating food and resources through hunting, scrounging, bartering, or gardening. In such a situation, a stash of firewood would provide enormous peace of mind.

The concept of apocalypse firewood is tied to the assumption that there will be a time when the gas runs out. The German word *versorgungssicherheit* evokes the fear of not having enough of something stocked up. When Russia cut off Germany's natural gas imports in 2022, for example, the demand for wood-burning stoves and *brennholz* (firewood) skyrocketed and the items became scarce and expensive.

When the gas runs out of my Husqvarna 550 XP, it is useless. If the gas is out in this apocalyptic scenario, the power is probably out as well, which means my lithium battery–charged EGO chainsaw is dead unless I can charge it via solar power or other means. Even if I could recharge it, the battery-powered saws are only good for less than an hour of steady cutting and the battery dies quicker when cutting into big wood and hardwood. Also, the lithium battery doesn't like the cold and craps out more than I'd like.

You may be familiar with the phrase, "Chop your own wood and it will warm you twice," which is often attributed to Henry Ford. He had the phrase carved into his mantel at his house in Dearborn, Michigan. But the phrase and the idea have been around longer.

Ford may have poached it from Henry David Thoreau's *Walden; or, Life in the Woods*, which was first published in 1854. "Every man looks at his wood-pile with a kind of affection," wrote Thoreau. "I loved to have mine before my window, and the more chips the better to remind me of my pleasing work. I had an old axe which nobody claimed, with which by spells in winter days, on the sunny side of the house, I played about the stumps which I had got out of my bean-field. As my driver prophesied when I was ploughing, they warmed me twice—once while I was splitting them, and again when they were on the fire, so that no fuel could give out more heat."

This sentiment is also expressed in the 1855 book *Thorn Cottage, a Poet's Home: A Memorial of Frederick Knight, Esq., of Rowley, Mass.*, in which Knight proclaimed: "He that cuts his own wood is twice warmed, he that earns his own bread is twice blest. The labor procures and sweetens the food. Let him try every other method—send to the Indies for condiments, and he will fail."

This idea also cropped up in other cold climates, as seen in Francis Kinloch's 1819 *Letters from Geneva and France: Written during a Residence of between Two and Three Years in Different Parts of Those Countries and Addressed to a Lady in Virginia*, where he observed the winter wood-gathering practices of the peasants in Jura, France: "The proverb of the country is, that wood warms a man twice."

I would like to call bullshit on this phrase.

First, there is the felling of the tree and getting it on the ground (warmed once). Then comes the limbing of the tree (warmed twice). Then the bucking, or cutting into usable lengths, of the tree trunk and bigger limbs (warmed thrice). After that, the wood needs to be loaded into a trailer or the bed of a pickup truck (warmed four times). (Even if you are dropping a tree in your backyard right next to your house, you still have to get your firewood off the ground and move it somewhere, maybe with a wheelbarrow or a sled.) Then, after the wood arrives at its destination, it must be unloaded (warmed five times). Once it is unloaded, the large rounds must be split (warmed six times) and then picked up off the ground and stacked (warmed seven times). Later, whether in large quantities or small, the firewood must be brought into the home for burning (warmed eight times).

The idea of firewood warming you twice is nonsense.

The only thing I can agree on is Thoreau's comment: "Every man looks at his wood-pile with a kind of affection. I loved to have mine before my window, and the more chips the better to remind me of my pleasing work."

Those who believe that chopping wood warms you twice are victims of what's called a "planning fallacy," predicting how much time it will take to complete a project with a view that is too optimistic. The result is an underestimation. Firewood harvest is also prone to Hofstadter's law: "It always takes longer than you expect, even when you take into account Hofstadter's law."

What's more, my example of being warmed eight times involves a gas-powered chainsaw, plenty of fuel, and a four-wheel-drive vehicle. In a world absent of gas and electrical power, the firewood would need to be harvested with axes, crosscut saws, and other labor-saving tools like cant hooks, log peaveys, and hookaroons.

Born in 1937 near Mountain, Wisconsin, the author Leon "Buckshot" Anderson gives a good description of the labor involved in felling and cutting a tree into lengths without the power of a modern chainsaw in his book, *Tall Timber, Tall Tales: Honoring the Lumbering Era and the Art of Tall Tale Telling*:

> I vividly recall watching my dad fall large pine trees with amazing accuracy. He'd eyeball the tree while it was standing and determine in which direction gravity would help the most in bringing it to earth. He'd grin at me and pound a short stake into the ground where he expected the crown of the tree to end up. Then he formed the notch with powerful swings of his double bit axe and finished the job by sawing towards the notch (back sawing) with his Swede saw. Rarely did he miss his stake by more than several feet and frequently he shattered the stake with a direct hit.
>
> Once the notch was completed the team of sawyers began cutting towards the notch from the opposite side of the tree with a huge two-man crosscut saw. This task required more energy than chopping the notch. Teams needed to establish a rhythm when using the crosscut saw while sawing. When one man pulled the saw his partner had to relax and not push the saw. This back-and-forth rhythm is not as easy to develop as one might think.
>
> In some cases, when completing the two tasks of notching and back cutting, four men shared the tasks necessary to make the tree

Many sawyers continued to use crosscut saws, like the one pictured here around 1880, even after prototype chainsaws had become available. This was partly because they considered the machines a threat to their occupation but also because early chainsaws were huge contraptions that were difficult to move through rough terrain. WHI IMAGE ID 3679

fall, two axe men and two saw men. Specialization such as this often took place in the big pine forests of yesteryear.

Once the tree was safely on the ground, another pair of workers took over while the sawyers moved on to their next victim. These were the buckers. The buckers also used two-man crosscut saws to saw the tree trunk into the required lengths. Most frequently the standard length of a log was sixteen feet but sometimes due to special orders, defects in the tree or 'crooks' in the trunk, log length could be shorter or longer.

The crosscut saw described here was once nicknamed the Swedish fiddle and the misery whip, with the latter hinting at the gruesome labor required to make it work. And Anderson's lengthy passage still fails to mention the job of limbing the tree. Once the tree was on the ground, an axe man would chop all the limbs off the trunk until it was a bare log, which the sawyers could then cut into lengths.

It took four men doing exhausting work to simply get a tree on the ground and cut into pieces. With a chainsaw, I can do the work of all four men in a fraction of that time, and I can do it more safely. And at the end of the process, rather than being exhausted, I usually have a pleasant warm buzz, as I do at the end of a good workout.

In an apocalyptic scenario, people unable to put up firewood by means of a chainsaw or an axe or a crosscut saw would be left harvesting scavenged wood, which is any wood that can be scrounged off the ground or, in the case of dead limbs, pulled off the lower trunk of a tree. The woodstove industry of today is entirely dependent on modern chainsaws.

In order to get the wood out of the forest without gas-powered vehicles or power equipment, a homesteader without draft horses—which is what the old-timers used—would have to skid out the firewood in sleds to the home or cabin. The logs could also be floated down a river—something I have done a few times with ash trees killed by the emerald ash borer. Ash trees like to grow in riparian zones, and I felled the ash into a small river on my homestead because I did not want to destroy the beneficial shoreline vegetation by hacking a trail for my ATV.

In Robert Gates's memoir, *What a Life!*, he describes growing up as a country boy in Wisconsin and being an early adopter of chainsaws. His worldview and business plan in the early 1950s was simple: there are those with chainsaws and those without chainsaws. "When I was young," he wrote, "I was just fascinated with chainsaws. They were just coming in at the time. Sometimes when I would be out hunting, I'd hear a chainsaw off in the woods someplace, and I'd hunt over that way, and then watch them with the chainsaw.

"At the time I worked at the Graber company in Middleton. When I got enough money ahead, I bought a chainsaw, and started to do some custom cutting at farms around. Most farmers didn't have chainsaws at that time, and they'd hire me to cut some firewood, or fence posts, or lumber. I did that [on] weekends and worked during the week at Graber's. Pretty soon, it seemed like everybody wanted me to come and cut something for them, and weekends didn't allow enough time, so I decided I'd make a profession of custom sawing, and I went at it full time. As it turned out, I worked in the woods and sawmills for forty-five years."

His definition of "custom sawing" was cutting wood for those who did not own a chainsaw. He was married with four young kids in 1952, and he was making enough money "custom sawing" to quit his day job. Thus is the power of the chainsaw. Thus was the demand for timber cut into useful lengths.

While building my writing cabin, I surfed the internet looking at images of cabins. My goal was to identify any construction techniques and design elements that might be useful. I perused images of rustic cabins, outdoor living, autumn colors, and cozy campfires. Collectively, this imagery is called cabincore, an aesthetic that romanticizes cabin life.

In these photos, I saw a lot of hipsters with axes, either splitting artisanal firewood or just looking tough in selvedge jeans with rolled cuffs and leather work boots that appeared to have never seen much work. Call it lumberjack live-action roleplaying. The axe seems to be the number one prop for poseurs trying to portray a vibe of rugged masculinity. Scrolling through thousands of these rustic images, I hardly saw any pictures of chainsaws, but axes were everywhere. How did hipsters develop this axe fetish?

It all started in Brooklyn, New York, in 2009. Peter Buchanan-Smith grew up on a farm in Canada and has fond memories of going camping and admiring his father's felling axe. Years later, as a designer living in New York City, he came up with the idea to paint the handles of axes with bright colors evoking summer camp and the Boy Scouts. His company, Best Made, became wildly successful.

On the company website, he recalled: "In our catalogs, we showed them being swung in the wild places where they belonged (even if some of our customers were only hanging them in their Tribeca apartments). We shipped thousands of axes to every conceivable corner of the world. Before Best Made, axes were most often relegated to a toolshed. Now they were hanging in art galleries, museums and the boardrooms of Fortune 500 companies."

Best Made turned into a lifestyle brand and sold other products. Buchanan-Smith eventually sold the company in 2019 and moved to upstate New York to write a book about axes. The result, *Buchanan-Smith's Axe Handbook*, is a wonderful book. The writing and research on axes are done with detail and care; the typography, photos, graphics, and charts are masterfully designed. And the physical book itself is a hardcover with a handsome heft.

In the book, Buchanan-Smith mentions chainsaws only a couple of times, as they pertain to the limitations of the axe: “If you plan on bucking a lot of logs then I’d suggest you consider the saw,” and “If you’re going to fell a tree, and you have good reason to do so, please don’t grab an axe. Grab a saw—preferably a crosscut saw or a chainsaw.”

For all of its faddish adoration, the axe is anachronistic in a world of chainsaws. My chainsaw does 95 percent of all my wood-cutting work. A pair of loppers takes care of the other 5 percent. The only exception is splitting firewood with a maul and splitting kindling with a hatchet. If the grid goes down, I hope to at least have some firewood precut into rounds with the chainsaw, because then it will be back to the axe. I have many axes but rarely use them. I remind myself of the importance of the axe with a line in Eric Sloane’s 1962 book, *Diary of an Early American Boy*: “The axe was the pioneer’s most important tool; a man could walk into the forest with nothing but his axe, yet fashion snares to catch game, fell trees, and fit them into a cabin. He could even clear brush for growing a garden and by holding an axe blade in his palm, he could use the sharp blade in the manner of a knife and whittle with it.”

In the early Bronze Age, axe heads had such value as a tool they were traded as currency and hoarded. The axe as described by Sloane is a heroic tool that shaped the American frontier. Now it is a semi-useful tool adopted by rustic hipsters to use as a prop in social media photos. But in the background of those photos are the straight cuts made by a chainsaw—an unsung marvel and a misunderstood machine that does all of the serious work with very little recognition. This book is for those who use chainsaws.

2

The Unnatural History of the Grid

The Saw as a Weapon against Darkness

I am fond of a quote that is often attributed to either George Orwell, Winston Churchill, or Rudyard Kipling: "We sleep soundly in our beds because rough men stand ready in the night to visit violence on those who would do us harm." It is actually from a 1993 *Washington Times* essay by Richard Genier, who was paraphrasing a similar insight from Orwell. I like to rephrase the quote as: "We live a life of comfort and convenience because rough men stand ready in the woods with chainsaws to prevent natural forces from plunging our society into darkness." Without men running chainsaws throughout the year, there would be no grid, and there would be no electrical power, and the modern world as we know it would come to an end.

"I think I shall never see, a poem as lovely as a tree," wrote Joyce Kilmer. I doubt Joyce ever observed a tree hitting a powerline and the destruction and danger it creates.

When I worked as a line-clearance tree trimmer for Asplundh Tree Expert LLC, the "rough men" I worked with came from all walks of life. Most gravitated to this vocation because they liked working outdoors and were mentally unfit to work in the confines of a cubicle or a factory. A few had college degrees, and some had studied forestry and other natural resource subjects. Some were barely old enough to drink while others had gray in their beards. One commonality that brought us together was the threats we all faced on the job. Every old-timer knew someone who had died. On the list of America's most dangerous occupations, logging and electrical line work are always near the top, and line-clearance tree trimming is a bastard hybrid of the two. The chainsaw is our primary tool.

A 2003 study by the National Institute for Occupational Safety and Health found that 15.6 percent of deaths among loggers were caused by machinery, and 84.4 percent were from being in the line of fire of falling trees or rolling logs. What made that line of work even more dangerous is that I worked with a wide variety of heavy machinery, all of which was dangerous to varying degrees, and the trees I felled were next to high-voltage lines.

During the early months of 2020, when the COVID-19 pandemic put the nation into panic mode, I found myself working with Mike French on the Y-180 line near the substation in Wild Rose, Wisconsin. The sixty-nine-kilovolt line was slated to be rebuilt to carry more electricity. The wood poles that held up the lines would be replaced with larger metal structures. When lines are replaced to carry more voltage, workers must expand the existing right-of-way. America's electrical grid is slowly getting upgraded, but many old lines are out there. I've come across ancient utility poles swallowed by the woods, stripped of wires long ago, their T-shaped posts still adorned with glass insulators. They seem out of place, like the lamppost in the forest of Narnia.

Our job that day was to clear a few more feet of trees from along the forest edge. In some spots we came across on the job, huge sections of mature trees needed to be removed, and where the line made a slight turn, the resulting dogleg section would be a full clear. In those places, our work resembled a logging operation. In other sections, where the landscape was more meadow than forest, only a few saplings needed to be cut down: easy money. Sometimes we were deep in the backcountry, and other times we were on the outskirts of town. Often, we worked in forsaken places, in overgrown ditches, and on the wild-grown side of train tracks, the kinds of spots where one might find a dead body in a Stephen King story. One of our surveyors once found a human skeleton still wearing a necklace and a ring.

The line rebuilds are plotted out years in advance and since the line was slated for a complete rebuild, it was pulled off a maintenance schedule and the right-of-way grew thick with brush. We were the shock troops, the trailblazers.

That afternoon, we came across a large red oak with a bad lean on a steep hill. An orange X was spray painted on the trunk, marking it as a hazard tree that could possibly hit the line. It had to be taken out. The first step was to get a rope tied into the upper branches so we could better control its fall.

I tossed the throw weight—a sixteen-ounce bean bag filled with lead shot—into the upper branches. It passed through a crotch on the main trunk and sailed deep into the woods. I put on my leather gloves, pulled in the throwline hand over hand, and coiled the line loosely into a folding cube—a nylon fabric box that folds down and pops up with a twist. When I'd pulled the throw bag just under the crotch where the line passed, I gave it a jerk to pop it over a branch and then finessed it to the ground on the front side of the trunk.

I threw the bomber line—a 150-foot three-strand rope we used for groundwork and rigging—over my shoulder. I removed the throw bag and looped the throwline around the tip of the bomber line with three half hitches. I walked back to the folding cube where the throwline was coiled and yanked in the line one armlength at a time. The bomber line climbed up the tree like a snake, went over the crotch, and came back to me. I untied the half hitches and walked the line back to the trunk of the tree. Then I tied a running bowline around the trunk, ran it up to the top of the tree, and cinched the line until it reached the crotch and tightened around the wood. I walked the line out in the direction I wanted the tree to fall.

Frenchy was down at the base of the hill in the Geo-Boy, a brush-cutting tractor that chews up brambles and turns small trees into shredded wood. It leaves behind a mowed field of woody stubble. After walking in it every day, one becomes untrippable.

He saw I had the rope stretched out and that the tree was ready to be felled. Per company rules, two men are required to fell trees of this size: one felling, one pulling the rope. And if it's possible that the tree might hit the line, one of the workers must be a foreman. Frenchy drove up, parked the mechanical beast, and walked over.

"So, what you got going on with this one?" he asked. Frenchy was my foreman. We were the same age, and he came from a family with a long history in tree work.

I pointed out a stick that I had stabbed into the snow where the rope was strung out. I had used the stick method to measure the height of the tree and mark where the top of the tree would hit the ground. The tree had the potential to hit the line, but if we felled it at an angle, we could drop it safely.

Frenchy looked up at the line, paced over to the stick, looked at the tree, looked back up at the line, back at the tree, back at the line. A line strike gets people fired. That's the best-case scenario. The worst case is death by electrocution, destroyed infrastructure, and power outages that could cause massive blackouts.

"Seems about right," he said.

I pointed out the head lean—a heavy forward lean puts tension on the upper trunk, which creates the danger of a tree splitting and kicking back when the faller is making the back cut. This forms what is called a barber chair, a partial tree stump with a large splinter forming the back of the chair. To prevent this, I would make a bore cut, which is used to safely fell trees with a strong lean.

"Sounds good," said Frenchy. "You have an escape route?"

"Over there," I said, "where the brush is cut out." Earlier, I brush-cut a path that angled back into the woods at a forty-five-degree angle from the rear of the tree.

"You the man with da plan," he said. "Let's flop it." He grabbed the rope near my marker stick and walked it out to a distance of one and a half times the height of the tree—the radius of the 360-degree area we called the danger zone.

I walked over to the base of the tree where I left my chainsaw and put on my saw chaps.

"Danger zone clear?" I yelled.

"All clear," hollered Frenchy.

I snapped down my earmuffs on my hard hat and yanked on the Husqvarna 372. As it warmed up, I knelt down beside the tree, rested the rear handle of the saw on my knee, and pressed the bar flat against the trunk. I scanned out to the edge of the right-of-way and picked the exact line where I wanted the tree to fall. I studied the felling sights on the top of my saw and adjusted my aim.

"Notch cut," I shouted over the rumble of the saw. Frenchy waved back as an acknowledgment.

I cut out an open-face notch in the oak and then yelled, "Making bore cut." Frenchy waved back again and tightened the rope.

The lower tip of the bar is called the attack; the opposite side is the kickback. I pressed the attack to the midsection of the trunk at full throttle until the cut was as wide as the bar. Then I rotated the saw until it was parallel to the notch and let the chain pull itself into the heartwood of the tree. Once the bar was fully through the trunk, I cut straight back, away from the notch, until a small strap of wood remained on the backside of the oak. The only thing keeping the tree upright was an inch of hinge wood at the apex of the notch and the back strap.

I studied the cuts. Everything was level. "Cutting the strap," I yelled at Frenchy. He waved and started reefing on the rope. I made an underlapping cut two inches below the bore cut and heard the crack. The trunk tipped forward. I wrist-snapped the chain brake and spun to my escape route. Four steps into the woods and a wait-a-minute vine tangled around my chaps and tripped me up. I had just enough time to throw the still-running saw away from me before I did a face-plant into the leaves and snow. I heard the thunderous crash behind me.

I got up and hit the kill switch on the saw. The tree came down where I wanted it, missing the line by an easy seven feet or so. I was peeved that I didn't see it fall, but it was satisfying nonetheless. More than satisfying. Exhilarating. Like knocking down an opponent with a left hook. Or feeling a jolt on the rod when a big brown trout hits a dry fly.

The feeling is more than just an adrenaline rush. I once read about a study of forager-farmers in the Amazon basin of Bolivia. A researcher compared the testosterone levels of the men before and after playing soccer and before and after felling trees by axe. Playing soccer increased their testosterone levels by 30 percent and cutting down trees boosted them by 48 percent. It's no surprise that Joe Louis, George Foreman, Gene Tunney, Muhammad Ali, and Floyd Mayweather chopped wood and felled trees as part of their training.

It made sense to me. There's a surge in the blood after felling a large tree. And to do it near a high-voltage power line made the whole thing more dangerous and intense. The stakes were much higher.

Frenchy waded into the canopy, found the running bowline, and untied it. He stepped back out of the canopy, pulled the rope out, and yelled, "Rope free!"—a signal to the faller that he was cleared to start limbing the tree.

Frenchy stretched out the rope to arm's length and coiled it into a couple of loops. Then he looked at his watch and threw the rope to the ground in a heap. "Screw this. Let's go take our break."

We hiked up the hill to the work truck. There was the fire station I set out earlier that morning. Late winter is the start of Wisconsin's wildfire season. On the way to work, I made note of a Smokey Bear sign with the "Fire Danger Today!" warning. The threat level was at high. As part of our job, we were trained in the basics of wildland fire prevention and suppression. The fire station consisted of an orange safety cone and a spade spiked into the ground next to it. Leaning against those two items were a Pulaski axe and a Macleod rake, forming a tepee of tools that were easy to grab. Next to that was a twenty-pound fire extinguisher and an Indian Smokechaser—a backpack bladder of water with a brass sprayer nozzle that operates like a pump-action shotgun.

Although it wasn't our job to fight forest fires, we were equipped to stamp out a small blaze before it got out of control. However, some crews in the region were sent to California to help with the wildfires. In the burned-over hill country, they worked long days dropping charred trees "into the black" while banking beaucoup overtime pay and per diem cash.

Ten yards from our truck was the JEL (job entry location) cone. On it hung a plastic pouch with the job briefing on a clipboard. Every morning, I filled out the briefing form, which listed the work entailed for this location and every possible hazard that could kill us or anyone who wandered onto our jobsite. After we finished the job briefing, we always performed our required stretching routine by the tailgate. We were considered "occupational athletes," and it is not a bad thing to get paid to do yoga.

Climbing trees is demanding and requires strength and mobility. Clearing low brush with a chainsaw feels like holding a twelve-pound kettlebell a foot off the ground and swinging it back and forth for hours. We also flipped logs off the right-of-way, and that action resembles the tire flipping you see in strongman events. Bucking and limbing trees requires squatting and bending the body at odd angles with a chainsaw in hand. And then there is the walking. Some days we went hot-spotting—basically a search-and-destroy mission. A helicopter crew flew the line and found hazard trees hidden along the edge. Once we had the location information, we geared up, not knowing what exactly we were in for. There was a disclaimer: "Maps have been developed from a desktop review of available data. Actual field conditions may vary."

I took off my hard hat for the first time that day, set it upside down on the truck's tailgate, and tossed in my safety glasses and gloves. I unbuckled my chaps and placed them under the covered bed.

"According to the new rules concerning the virus, we're not supposed to take breaks together," I said. "Not that I particularly give a shit."

"Jesus, we work together," said Frenchy. "If you get it, I get it. There is no way we can get anything done with these new rules. We will spend half the day sanitizing the entire goddamn truck and all the tools. It's pointless."

He picked up the jug of bar oil on the tailgate. "You touch it, I touch it."

He picked up the metal can of saw gas. "You touch it, I touch it."

He picked up the scrench, an essential tool that is half screwdriver, half wrench, used for messing with chainsaws. "You use it, and then I use it. And we get the virus and die. Next thing you know, they're gonna have us wear masks. Screw it. Let's take a break."

As a permanently mobile crew, our work life in remote areas revolved around our four-wheel-drive work trucks. Everything was based out of the trucks, which were usually either Chevy Silverados or Ford F-250 Super Dutys. We ate, answered emails, took breaks, watched safety videos on tablets, and filled out paperwork—all in the trucks. It was where we took shelter during lightning strikes, heavy downpours, and cut-you-down sleet storms. All crew members had commercial driver's licenses. We were a mobile unit that went where the line went, and we hauled heavy equipment up and down the line and dropped it off at select staging areas. We had an armada of semis, pull trucks, fuel trucks, bucket trucks, wood chippers, trailers, minidumps, tractors, UTVs, skid steers, mowers, Skytrimmers, processors, feller-bunchers, Argos, and Marsh Masters, but most of the time, we could be found in the work trucks.

We climbed into the cab, and Frenchy fired up the heater. He dug out slabs of cold pizza from his lunch box, and I opened up my thermos filled with venison chili. Hillbilly, our general foreman, had stopped by the day before. He was a former Halliburton mercenary who abused plenty of booze and weed in his day. He got cleaned up and has a couple of kids and a wife who is a university professor. An all-around good guy. He has the nickname Hillbilly because he grew up in rural Missouri and has many redneck stories to tell.

He'd given us each a paper titled "Critical Infrastructure Work Release." We were supposed to keep it in our personal vehicles during the lockdown. The key paragraph: "Our employees in region 062 support various utility providers' critical infrastructure throughout the states of Wisconsin and Michigan to ensure the delivery of safe and reliable power to their customers. Our employees primarily provide line-clearance tree trimming/removal; however, some of our staff provide the management and logistical operational support for these activities."

Hillbilly gave us a plastic spray bottle of sanitizer solution—a homemade cocktail of isopropyl alcohol, hydrogen peroxide, and water that he mixed himself. It was the best he could do, he explained. Everything was sold out. Also, he brought with him more rules that the upper management cobbled together concerning the virus: Take breaks separately. Sanitize all tools before use. Sanitize the cab and dashboard before driving. No more than two people per truck. Maintain social distance despite having to work, ride, and drive together.

"Jets," said Frenchy through a mouthful of pepperoni. Being parked on top of the hill gave us a wide view of the horizon. Two F-16 fighter jets streaked through the clouds leaving contrails. Since we worked on trees, we were always looking up into the air. We hadn't seen any planes for days. Commercial aviation had come to a halt during the apex of the pandemic. Now only military aircraft took to the air—and for reasons unknown to us.

"Probably from Volk Field," I said, referring to the Air National Guard base over in Juneau County. It was a couple of hours away by car. "They're scrambling their jets. We pissed off China. We're on the cusp of World War III."

"Man, this is messed up. What is going on with this world?" Frenchy thumbed his phone, scanning headlines on his Facebook feed, his only source of news.

I thought back to late January 2020 when I had gone to work feeling a little sick. It was minus ten degrees, and I was bundled in layers of cold-weather work wear: merino wool, Carhartt canvas, polypropylene long johns, 1,200-gram ice fishing boots. Down the right-of-way was a sapling-covered rock pile that needed to be cut out by hand. In this part of

Wisconsin, the glaciers from the last ice age deposited rocks and boulders throughout the countryside, and farmers have been plucking them out of their fields ever since. Right-of-ways are easy places to dump the rocks. Years pass, and saplings grow through the gaps in the piles.

The Husqvarna 372 chainsaw felt like a boat anchor. My arms were noodles. I was burning up yet freezing at the same time. Fever with chills. It was subzero weather, and I was overheating. I took off a layer and put it back on. My chest felt compressed and tight, and each raspy breath brought in needles of icy air that penetrated every cell in my lungs, making them feel hard and frozen. It was a kind of sickness I'd never felt before. I went home.

I was healthy after a couple of days of bedridden hell. I needed a doctor's note saying I was fit to return to work. At the clinic, I asked the doc about a long shot: I heard in the news that this contagious virus from China had made landfall on the West Coast and was spreading. Could this be what made me sick? Could it possibly have made it to Wisconsin already? She waved off my concerns. "No," she said, "this is just a flu bug going around."

I reflected for a moment on what could go wrong at that point in time. We were a society at the start of a pandemic. A tree could hit a power line, causing a massive blackout across a vast region. Winter still lingered and warm spring weather hadn't yet arrived. Furnaces would shut down and homes would grow cold. Darkness would set in. People would panic-buy batteries, propane, and other supplies until the shelves were bare. Wildfires could spread across the landscape, burning down lightless homes. Firefighters, law enforcement, and other first responders would struggle to do their jobs without electrical power. Hospitals would fill with patients suffering from the new virus and function at half capacity as backup generators struggled to life after sitting idle for years. As elderly patients suffocated and died under airless ventilators, TikTok nurses would flash dance against the emergency strobe lights to film their final upload.

The grid can go down faster than you think. If you happened to be standing in a right-of-way in Cuyahoga County in northeast Ohio on August 14, 2003, you might have noticed a 345-kilovolt transmission line starting to sag. If you were there at 2:02 in the afternoon, you would have seen a tree come in contact with the line. At 3:05, on another right-of-way in Parma, Ohio, another 345-kilovolt transmission line made contact with another tree. At 3:32, another transmission line called the Hanna-Juniper interconnection sagged into a tree and a cascade of failure rushed ahead: circuit breakers were tripped and lines crashed one after another. Alarm and monitoring systems failed. Communication breakdowns occurred as controllers tried to figure out what was going on. More lines failed throughout Ohio, and power surged across state borders, instantly overloading other lines.

By 4:13, power was down in New York, New Jersey, Maryland, Connecticut, Massachusetts, Michigan (including the Upper Peninsula), Ohio, Pennsylvania, and Ontario. A few pockets in this area remained unaffected, but it was a Black Sky event for an estimated fifty-five million people. It was the end of their illuminated and electric-powered world and a chance to see the Milky Way.

One year later, an investigative task force report listed the four main causes of the blackout. Number three was that FirstEnergy, the Ohio electric utility, "failed to manage adequately tree growth in its transmission rights-of-way."

A skilled chainsaw operator could have prevented the biggest blackout in American history with three simple cuts: two for the notch and one for the back cut.

At the end of our break, Frenchy headed back to the Geo-Boy, and I grabbed a roll of toilet paper from my kit bag. I shit in the woods more than most bears. According to our contract, we are allowed to take the work truck to the nearest gas station to use the restroom, but in the rural areas where we work, the nearest convenience store could be a long way off. It's sometimes easier to just go in the woods.

Angled off the right-of-way, I found a private spot by some young white pines. I looked over the progress we made on the right-of-way so far. The Spanish architect Antoni Gaudí said, "There are no straight lines or sharp corners in nature." That is, until line-clearance tree trimmers come along. You never see power lines in landscape paintings. People would rather paint and view renditions of vistas unaltered by mankind. However, I've come to view the utility lines and the right-of-way land they stretch out upon as elegant and essential. They are the thin fragile lines that separate us from modern civilization and the Dark Ages, or at least preelectrical

The most common threat to electrical energy distribution is a tree making contact with an energized line. Because the width of a right-of-way is determined by the voltage carried by the lines, workers with chainsaws must create giant corridors through the forest for large transmission lines.

America. Without these suspended wires providing us with precious energy, our civilization would quickly collapse. The National Academy of Engineering considers the grid to be the greatest engineering achievement of the twentieth century, and yet it is something that we collectively take for granted as a nation. For me, the utility line is a scrimmage line. It is the most valuable real estate in the US worth fighting for. And chainsaws are our greatest weapons in this battle.

Line-clearance workers commonly use the expression "ground to sky," meaning that when you stand next to a wooden clearance limit stake topped with high-vis pink ribbon, you should be able to look down the right-of-way and see the next stake and the next one and the next one. From ground to sky, not a single branch or twig should be sticking out past that invisible line. It should look as if the hand of God came down with a meat cleaver and sliced the forest in half. Essentially, they are making a hallway through the woods. Everything at ground level is cut flush to the stake-to-stake sightline. If a branch is poking out forty feet off the ground, they climb the tree and cut the limb with a handsaw or a small chainsaw called a climb saw. If the branch is lower and reachable, they cut it down with a polesaw.

While pruning near energized lines, they use the guidelines set forth by Dr. Alex L. Shigo, a former chief scientist with the US Forest Service who wrote a small but insightful book on the subject. They work by the 90-3-90 concept: 90 percent of the time, only three branches need to be removed to provide 90 percent of clearance. All of their cuts are at the nodes to prevent excessive sprouting and to prevent rot and cracks. The first cut is simply to drop the branch from the tree, which leaves a small stub connected to the trunk. From there, they cut as close as possible to the branch collar while taking care not to wound the trunk. Each collar is different, and every cut requires some study. If the cut is properly made, donut-shaped tissue will form around the cut and the tree will heal itself.

For each tree, they have to make a mental calculation of not just which branches to take but how their removal will affect the other branches on the rest of the tree. All future growth needs to grow away from the lines. And for a while it will, but eventually the tree will send sprouts back into the right-of-way to capture the open and abundant sunlight. Trees want to grow toward the open sunlight where the wires are; they grow in a way that is a threat to human civilization.

In the process of removing trees that need to be removed, they harvest the logs on behalf of the landowners. If the logs are large, they use a grapple mounted on a Bobcat skid steer, if one is available on that jobsite. If not, they hump it to the edge of the woods by muscle and teamwork and grunting. The grapple on the skid steer is claw-controlled by hydraulics and can snatch and lift massive pieces of wood. They pile the wood off to the side for the landowner to sell for pulp or lumber, cut into firewood, or mill into boards.

From what my colleagues and I have observed, the majority of the wood left behind is never touched, which is a sad waste of a natural resource. Wood is a sustainable energy source. A based woodsman will realize that a pile of wood left behind on a right-of-way is a labor-saving bargain since most of the hard work has been done. Besides that, the wood is located in

An old saying goes: "Trees don't heal, they seal." When a tree trimmer makes a pruning cut with the chainsaw bar aligned with the trunk of the tree, callus tissue forms a symmetrical donut shape around the wound—a sign of pruning work done well.

a recently cleared right-of-way, easy to get to by truck, tractor, or ATV. All that is left is cutting it into firewood-sized chunks, hauling it off, and then splitting and stacking it at leisure.

One exception to all of the careful, surgical cuts that workers make on a tree is mechanical pruning. This is usually accomplished by the Skytrim, a machine that is tracked like a tank and about the size of one. The operator sits up high in a cab, controlling a telescopic boom that can reach out to the treetops. At the end of the boom is a circular saw about the diameter of a manhole cover. Although the operator can adjust the cutting angle by tilting the cab and the boom side to side, the cuts are imperfect, and small branch stubs are left sticking out from the tree. This is considered an acceptable trade-off as the Skytrim can prune vast sections of the right-of-way much faster than a climber or a guy in a bucket making cuts. Usually, the Skytrim is brought in for long sections of rough-terrain forests out in the middle of nowhere.

Anything that holds up an energized line is called a structure. It doesn't matter if it is a wood pole, a tubular steel pole, or a giant, latticed, self-supporting tower made of galvanized steel—it is a structure. The spaces between the structures are called spans. The spans vary in length depending on the voltage of the lines, the size of the structures, and the topography of the land. On the transmission lines that I worked on, spans were around a hundred yards or more—about the length of a football field. We used them as a unit of measurement, as in: "We had to hike four spans to cut out a dogwood swamp."

Stay in this line of work long enough and you develop the ten-span stare. It is like the thousand-yard stare of combat vets who have an emotionally detached gaze after enduring the horrors of war. With a ten-span stare, you look back at a cleared transmission line strung out far to the horizon line and you remember the sweat, blood, diesel, bar oil, mixed gas, hydraulic fluid, arguments, horsepower, and hours invested into forcing that straight-line path through the natural world. The cost is great.

When we are done, the narrow strip of land looks like it was ravaged by the Once-ler, the character from the Dr. Seuss book *The Lorax*, which is a fable on environmental degradation. In this children's book classic, the Once-ler comes across an area of Truffula Trees and proceeds to cut them all down to make Thneeds to sell for profit. As he expands his operations, he destroys the forest of Truffula Trees, which in turn affects the Bar-ba-loots, small furry creatures that use the trees as a food source. Also, his Thneed factory pollutes the air and water. In the end, the Once-ler becomes the victim of his own greed by cutting down the last of the Truffula Trees. He can no longer make Thneeds and is left to live alone in an ecologically devastated wasteland.

The only thing we have in common with the Once-ler is that we completely clear the right-of-way land—when we're done, not a single woody stem grows from the ground. However, the Once-ler is motivated by capitalistic greed; we are motivated by our mission to protect the infrastructure. After all, the greatest enemy to our electricity-dependent way of life is not necessarily a cyber- or terrorist attack, a coronal mass ejection, or an electromagnetic pulse. The threat comes from Mother Nature herself—and our own neglect.

The Truffula stumps in *The Lorax* do not regenerate on their own through suckering, and at the end of the story, a boy is given a Truffula Seed—the last one of all—to regrow the forest. This is not how things proceed on the right-of-ways. Nature abhors a vacuum, and the land reverts back into a woody jungle in a few years. Some trees do not grow back after being cut down, while others send up suckers with renewed vigor. Seeds that lay dormant in the soil for years can now spring to life in the new sunlight. From a business perspective, line-clearance work is a self-licking ice-cream cone. As long as sun shines upon planet Earth, there will be money to be made.

Some sections of a right-of-way host human activity, but most do not. The most common remnants of human activity that I saw were hunting blinds and tree stands, both maintained and collapsed. Sometimes I saw four-wheeler and snowmobile trails passing through the brush. Marijuana plots were not uncommon. Occasionally, I came across a vegetable garden. Right-of-ways are also popular dumping grounds for car tires, deer carcasses, ancient farm implements, and junk automobiles. And I regularly came across the detritus of beer cans, shotgun shells, and used condoms. With the exception of these artifacts, the landscape is in a perpetual state of rewilding itself.

In his 2007 book, *The World without Us*, Alan Weisman envisions a world in which humans have suddenly vanished from the earth. He describes Manhattan transformed by floodwater, decay, weeds, trees, wildlife, and rust: "Gutters buried under tons of leaf litter provide new, fertile ground for native oaks and maples from city parks. Arriving black locust and autumn olive shrubs fix nitrogen, allowing sunflowers, bluestem, and white snakeroot to move in along with apple trees, their seeds expelled by proliferating birds. . . . Like a gap in the forest when a giant tree falls, new growth will rush in. Gradually, the asphalt jungle will give way to a real one."

The world without us is the world I worked in every day, and line-clearance workers are the ones who make the gaps in the forest. Right-of-ways are put on multiyear pruning cycles, but everything grows back—thicker and nastier. The land is cleared of woody vegetation, hazard trees are removed, and sometimes Garlon 4 Ultra is sprayed on the stumps (all of us were state-licensed herbicide applicators). And then it is left alone.

When my fellow workers and I returned years later, chainsaws snarling, trees as thick as our forearms now choked the landscape. Deer trails wound through the new growth. Seedlings were everywhere. I saw the weathered stumps cut by tree trimmer crews that had passed through decades before me. Some of the stumps were ancient, and the chainsaws used to cut those trees must have been ancient as well—heavy and cumbersome compared to the sleek, quick-handling pro-grade saws we use today.

The mature trees along the edge of the right-of-way had new branches reaching toward the line and drinking in the abundant sunlight. And then we hacked it all down again: Cry "Havoc," and let slip the saws of war! It's a fight against phototropism. It's a war against nature. This is the unnatural history of the grid—a place of never-ending destruction and renewal.

William James, the late-nineteenth-century psychologist and philosopher, suggested waging a "war against nature" as a way to gain the virtues that emerge from battle (such as heroism, courage, and grit) without actually having to go to war against other people. Just battle nature instead. He called it the "moral equivalent of war." "Human life with no use for hardihood would be contemptible," he wrote. "Our ancestors have bred pugnacity into our bone and marrow, and thousands of years of peace won't breed it out of us." His vision was something like the Depression-era Civilian Conservation Corps, which employed young men to clear forests for campgrounds, build roads and trails, fight fires and floods, plant trees, and improve trout streams.

In Cormac McCarthy's novel *The Road*, a father and son walk down a highway pushing a shopping cart through a dystopian landscape: "He pushed the cart and both he and the boy carried knapsacks. In the knapsacks were essential things. In case they had to abandon the cart and make a run for it. Clamped to the handle of the cart was a chrome motorcycle mirror that he used to watch the road behind them."

In such an apocalyptic world, I would follow the right-of-ways that chainsaws have carved out of the landscape. The roads would be magnets for danger, but the right-of-ways run through wild country and farmland. They are the overlooked and uncelebrated paths through the woods that lead in and out of all towns and cities.

After work on that day in late winter 2020, I drove home with my son after Little League practice. I pointed to a transmission line poking above the treetops in the distance. "See that line? That's the A-105," I said. "It's the big high-line of this area. If you are ever lost and need to get home, you follow that line. It will lead you right to our land, right to our turkey-hunting spot, and from there you can pick up the trail that goes back to the house. Follow that, and you will be safe."

3

Something Sharp This Way Comes

The Saw as an Ever-Evolving Design

There have been two milestones in the development of the modern chainsaw. One was the relentless mechanical refinement by the German engineer Andreas Stihl, which reduced the chainsaw to a device that could be handled easily by one man. The other was the work of the American logger Joseph Cox, who took his observations in the natural world and revolutionized the design of saw chains. The companies these men founded still dominate the chainsaw world today. One market research firm estimated the global chainsaw market to be $3.85 billion in 2021, and they say it will continue to grow. North America makes up the biggest share of that market.

Ask anyone in the world to draw a simple sketch of a saw, and it is very likely they will draw the teeth of the saw as a line of side-by-side triangles. Yet, this type of saw was abandoned in 1876 when Henry Disston & Sons, the largest saw maker in the world, invented a saw tooth that did not cut. It cleaned, instead. Later, these teeth would be referred to as rakers. The rakers clean sawdust out of the kerf—a slit made by a saw or other cutting tool. The kerf is the empty space left behind when a material has been cut. Think of it this way: a razor blade makes a narrow kerf, a Bowie knife makes a wide kerf.

In *Saws That Sing: A Guide to Using Crosscut Saws*, a handbook written in 2004 for the US Forest Service, David E. Michael explains, "Wet or green wood is hard to remove from the kerf because it is resilient. Even when the fiber is dislodged, it clogs a saw's cutter teeth. A special kind of tooth, the raker, allows the cutter teeth to work more effectively with less effort. Even though the rakers don't sever fiber, they perform the other two functions of saw teeth: breaking loose the cut fiber and removing it from the log. Rakers remove material whether the saw is being pushed or pulled."

The innovation of the rakers in the late 1800s made the saw much easier to use and spurred more development and experimentation in the shape, size, and spacing of rakers. The gullet—the empty space between the rakers and the teeth—is where sawdust collects as a saw cuts through wood. The gullet must be large enough to hold the sawdust until the saw is pulled away from the wood and the sawdust can fall away. Saws without rakers make sawdust, not wood shavings or chips. They also cut slowly and require more labor.

The raker tooth was a breakthrough in the world of cutting wood, and the concept would later be applied to chainsaws. "After 1890, most trees were being sawed down by the improved crosscut, not chopped down [by an axe]," wrote Malcolm Rosholt in *Wisconsin Logging Book 1839–1939*. "And within the next four decades, most of the good timber in the forest was on the way to a sawmill. In short, the crosscut saw, with a raker, had a lifespan of about sixty years before it was replaced with a practical chain saw that one man could handle, about 1950."

In the early twentieth century, tinkerers around the world were pursuing the concept of an "endless saw," a type of machine-powered saw that would continue to cut long after human arms collapsed from pushing and pulling the misery whip. An early idea was the "drag saw," which involved attaching a crosscut saw to a mechanized contraption to do the back-and-forth cutting work. But in later visions of the "endless saw," the saw did not cut back and forth but rather traveled in a circle—endlessly whirling with cutting teeth.

One odd footnote in the history of the chainsaw was a development by the Scottish physicians John Aitken and James Jeffray between 1783 and 1785. Their chainsaw was used for cutting diseased bones and performing symphysiotomies, an obsolete surgical procedure where the cartilage in the pubic area is cut to open up the pelvis during a difficult childbirth. It had a hand crank, like an eggbeater, that caused a chain to revolve around a small, pointed bar. The German doctor Bernhard Heine further developed the idea in 1830, and it became known as the chain osteotome—*osteotome* being the medical term for a bone-cutting tool.

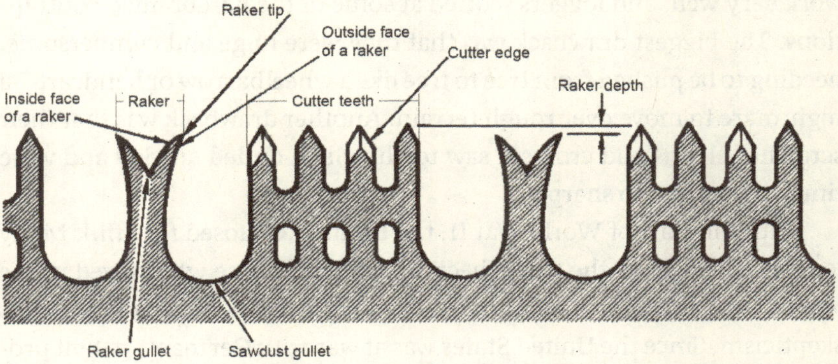

The rakers and sawdust gullets of early crosscut saws would later be applied to chainsaws in the form of depth gauges (also called rakers). The depth of a cut is determined by the height of the rakers. USDA FOREST SERVICE *CROSSCUT SAW MANUAL*

In 1858, the first US chainsaw patent was awarded to Harvey Brown for a sectional band-saw-like device. George Kammerl was awarded a patent in 1863 improving on this idea, describing his creation as "a saw blade that is formed of an endless chain . . . moving continuously in one single direction." Neither of these ideas worked, but they planted the seed for other innovations. Many Rube Goldberg–esque contraptions were made in the next half century. Some of them worked, many did not, but they all had one thing in common: they were heavy and bulky and not practical for fieldwork whatsoever.

The breakthrough came from Charles Wolf in 1920 with his Wolf Electric Drive Link Saw. The saw was powered by an electric motor and ran off a portable generator. He designed three different models, all of which weighed between seventy and ninety pounds. Meanwhile in the Black Forest of Germany, Andreas Stihl was testing his invention of a two-man chainsaw with a gas engine. And at the same time on Mount Dolmar in the Thüringian Forest, Emil Lerp invented a gas-powered two-man saw. His chainsaw company would later be named after the mountain. By 1930, Stihl had developed a saw that could be handled by one man. In 1936, he refined his ideas down to a forty-six-pound chainsaw, which was exported to North America and garnered interest through promotional demonstrations.

By the 1940s, chainsaws were becoming more prevalent in the American woods, and loggers were the guinea pigs. Most early models did not work very well, and loggers scoffed at some of the harebrained contraptions. The biggest drawback was that they were huge and cumbersome, needing to be pushed from tree to tree like a wheelbarrow or handcart—a nightmare to move over rough terrain. Another drawback was that their scratcher chains had crosscut saw teeth, which dulled quickly and were time-consuming to sharpen.

With the start of World War II, the US market closed for Stihl. Many loggers throughout the world became soldiers. Those who stayed in the woods were now shorthanded and looked at the mechanical saws with less skepticism. Since the United States was at war with Germany, patent protections were tossed out the window, and American manufacturers began producing gas-powered saws based on Stihl's design. Andreas Stihl was a member of the Nazi Party and ran his company, A. Stihl Maschinenfabrik, throughout the war despite his factory being bombed. After the war, Allied forces deemed him a *mitläufer*, a term for people who did not believe in Nazi ideology but were compelled or obliged to follow along because of political peer pressure.

Before, during, and after both world wars, small improvements were made to chainsaws, but many loggers still used their crosscut saws. All inefficiencies have their patrons, and one reason many loggers did not quickly warm to the early chainsaws—besides the fact that they were unwieldy, clunky contraptions—is that they believed it would put them out of a job. It's an old and understandable sentiment that still exists today when a new technology disrupts an industry.

The new technology that disrupted the chainsaw industry came from a man who was seemingly more curious than his peers. Joseph Buford Cox was a jack-of-all-trades with a fifth-grade education. While working as a logger in Oregon in 1946, he tested an early chainsaw powered by a motorcycle motor and pushed around in a cart. He was unimpressed by the way it cut with its scratcher chain that mimicked crosscut saws. With this on his mind, he noticed how the larvae of pine sawyer beetles (*Ergates spiculatus*) quickly chewed through an old stump and left a trail of sawdust. Cox studied them under a magnifying glass and noted how the curved jaws of the larvae tore into the wood fibers from side to side instead of straight ahead. The larvae were able to go through the wood by digging, not scratching the wood fibers. They produced minuscule wood chips, not sawdust.

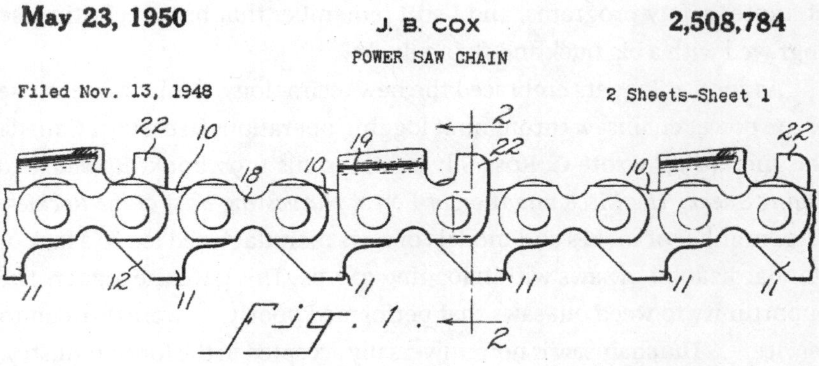

Except for some small tweaks and refinements, the chain invented and patented by Joseph Cox in the late 1940s is the universal design still in use today. GOOGLE PATENTS

Think of it this way: you are tasked with digging a narrow trench. You start out with a sharp stick (scratcher chain) and rake it again and again in the same spot and you eventually gouge a line in the earth. Over yonder, you see a Warren hoe (a hoe with a pointed tip) leaning against the barn. You give this hoe (the larvae-inspired chain conceived by Cox) a try and notice you are able to bite into the earth at an angle and dig out a trench faster and far more efficiently. Instead of making small pilings of dirt with the stick, you are now flinging big clumps of dirt with the curved and pointed Warren hoe.

Cox applied his observations to the design of a new saw chain with alternating curved teeth angled to the left and right to mimic the boring movements of the larvae. He patented the idea, and in 1947, he formed the Oregon Saw Chain Manufacturing Corporation. Cox assembled his product in the basement of his Portland home, and within a few years, his "bug chains" dominated the global market. He changed the company name to Omark (meaning Oregon mark) to make it easier for foreign buyers to pronounce. By the 1980s, the company had thousands of employees, and it expanded into other markets, acquiring Outers Laboratories—maker of Gunslick gun-cleaning products based in Onalaska, Wisconsin—in 1981. On a personal note, if it wasn't for Joseph Cox's company, I never would have received something I greatly treasured as a twelve-year-old boy. Outers Omark generously donated gun-cleaning kits to young graduates of hunter safety programs, and I still remember that brown plastic case engraved with a big buck and flying ducks.

"As soon as loggers embraced the new technology, the increase in use of the power chainsaw throughout logging operations in eastern Canada was meteoric," wrote C. Ross Silversides in his 1997 book, *Broadaxe to Flying Shear: The Mechanization of Forest Harvesting East of the Rockies*. "The number of makes and models of saws manufactured led to a lack of standardization of saws within logging camps. This gave the loggers the opportunity to weed out saws that performed poorly, or were difficult to service.... The chainsaw is now universally accepted in the forest industry. Bucksaws have completely disappeared, and axes have almost disappeared as logging tools."

Not only did the chainsaw revolutionize the logging industry, it also revolutionized rural life. As Silversides noted, "It was portable into other sectors of the [loggers'] own economy. The power saw had wide use around a farm or rural dwelling in the production of fuelwood, fenceposts, etc. Having these alternative supplementary applications to the main function of logging made the ownership of a power saw attractive." The postwar years were a frenzied race by many manufacturers to dominate the chainsaw market. It was more than just a cottage industry serving forestry workers. It was like the start-up culture in Silicon Valley, but instead of a cluster of innovators in one location, manufacturers of chainsaws and related gadgets popped up around the world. When software developers realized that computers were not just for scientists and that personal computers could be utilized in every home, the market and money-making possibilities exploded. It was the same for chainsaws: they weren't just for lumberjacks anymore.

One maverick of the chainsaw world was Elmer C. "Carl" Kiekhaefer, a farm boy who studied electrical engineering at the Milwaukee School of Engineering and the University of Wisconsin in Madison. In 1939, when he was thirty years old, he purchased an ailing farm implement manufacturer that was saddled with three hundred defective boat motors. Kiekhaefer had planned to invent milk separators for the dairy industry, but instead, he redesigned the junk boat motors, sold them, and received orders for more. Kiekhaefer became a moving force in the boating industry, eventually moving his thriving business from Cedarburg to Fond du Lac and calling it Kiekhaefer Mercury, which was later changed to Mercury Marine. He continually refined his motors, putting special emphasis on making them streamlined, slim, and compact—the same Holy Grail that entrepreneurs throughout the fledgling chainsaw industry were pursuing.

During World War II, Kiekhaefer turned his attention to chainsaws and won a government bid with a design to produce air-cooled, two-man chainsaws that used Mercury motors. Like Stihl, which built two-man saws for Germany's Nazi army, the Kiekhaefer Mercury company became one of the world's largest chainsaw makers during the war. Mercury produced the motors for the saws, while Reed Prentice Corporation made the bars and chains. Mercury also partnered with Henry Disston & Sons to produce chainsaws during the war. After the war, Kiekhaefer turned his attention back to outboard motors. With the return of American soldiers and the resulting baby boom, the US economy took off. Kiekhaefer became a titan in the boating industry as interest in recreation and water sports exploded across the nation.

The forty years after the end of World War II became the Wild West era of chainsaw design and development. Many models and manufacturers came and went. McCulloch and Homelite dominated the US market for many years. Husqvarna in Sweden had been founded in 1620 as a rifle maker and went on to manufacture everything from sewing machines to typewriters to dirt bikes, and as the chainsaw market grew, this manufacturing powerhouse entered the fray in 1959 with its A90 prototype. It has dominated much of the market ever since. To mark its sixtieth anniversary as a chainsaw manufacturer, Husqvarna produced a greatest hits album composed of twelve tracks. Each track is the sound of one of twelve different models of chainsaws the company has manufactured over the decades. The vinyl LP is extremely rare, but the audio can be found online. Just as Husqvarna released its first model in 1959, Stihl produced the Contra/Lightning, its first direct-drive saw, and that model alone established the company as a global leader.

The chainsaw became better known to the general public throughout the second half of the twentieth century. In 1952, the first issue of *Chain Saw Age* was published, and the magazine lasted until 1992. McCulloch was the first saw maker to advertise on national television, in 1961. By the 1980s, the chainsaw settled into a form and function that resembles

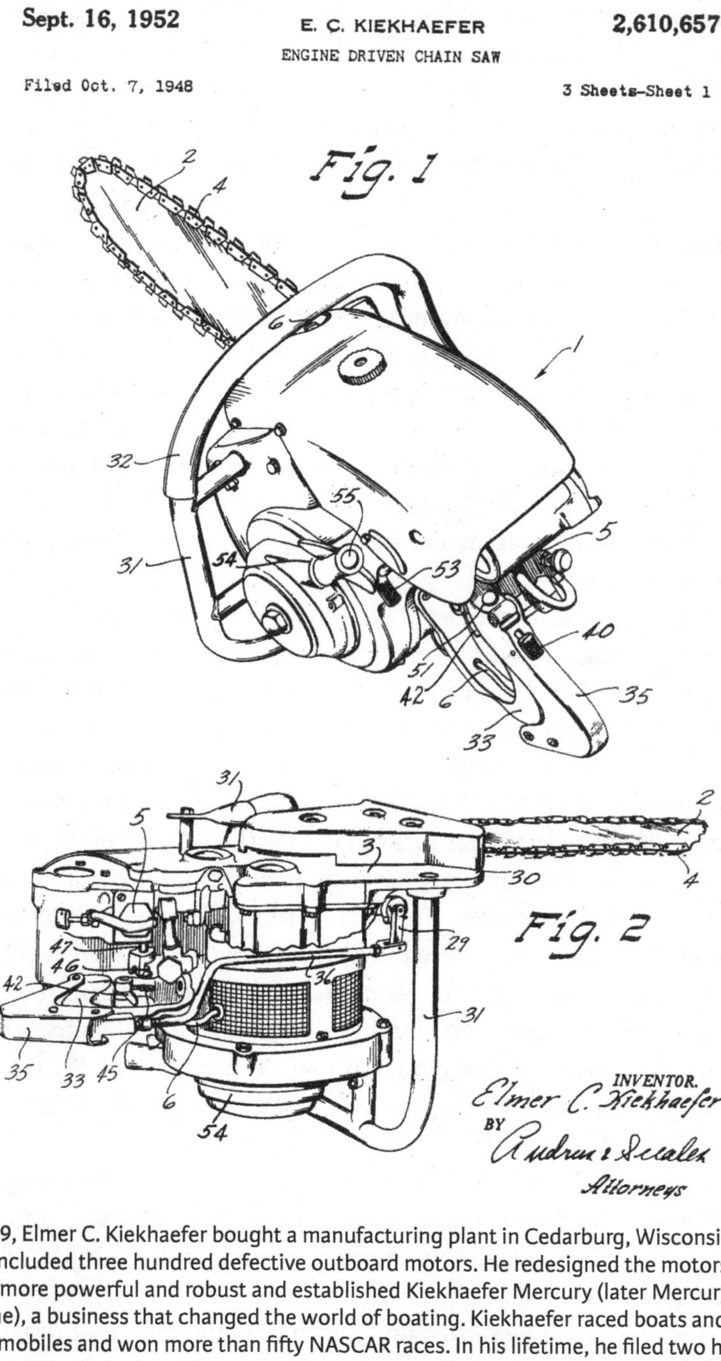

In 1939, Elmer C. Kiekhaefer bought a manufacturing plant in Cedarburg, Wisconsin, that included three hundred defective outboard motors. He redesigned the motors to be more powerful and robust and established Kiekhaefer Mercury (later Mercury Marine), a business that changed the world of boating. Kiekhaefer raced boats and snowmobiles and won more than fifty NASCAR races. In his lifetime, he filed two hundred patents, including this engine-driven chainsaw design in 1948. GOOGLE PATENTS the chainsaws of today. The groundbreaking work is over, and now the advancements being made are all about refinement. The best resource on the historical evolution of chainsaws is *Chainsaws: A History*, written by David Lee with Mike Acres. Acres is the founder of the Chainsaw Collector's Corner, the online epicenter for information about antique chainsaws. The book is extremely detailed and well researched and is loaded with color photos—a must-have for those who appreciate vintage saws.

The Tool and Its Parts

As with seat belts and shatterproof glass in automotive safety, it took some time (and blood loss) for chainsaw manufacturers to incorporate safety features into their machines. The most important mechanism is the chain brake, which instantly stops the chain from spinning. On most saws, this is a plastic fin that acts as a guard for the left hand. The operator pushes the guard forward to brake the chain and pulls back to release it. Engaging the brake can be done with a wrist snap movement without the left hand needing to let go of the handle. The chain is stopped by a spring-loaded steel band wrapped around the clutch drum. The brake can sometimes—but not all the time—prevent kickback if, when the saw jerks upward, the brake/hand guard is bumped during that split-second movement.

One always engages the chain brake when starting a saw and when walking from cut to cut. This is extremely important when limbing a tree. When some trees fall to the ground, they end up in a tangled jungle of branches that a sawyer must wade into. There can be many tripping hazards, and after a few cuts, a sawyer may end up waist-deep in branches that must be cleared out. With a flick of the wrist to engage the chain brake, a sawyer can hold the now-safe saw in one hand and fling branches that are blocking the way with the other. When hiking a short distance to another tree, one should engage the chain brake, since there is no point in leaving the chain idling. A good habit is for a sawyer to set the chain brake every time they need to take a few steps to get into position for the next cut. Before the creation of the chain brake, the slightest trip, slip, or fall could land a sawyer upon their own running saw. Another safety device to counteract this is the centrifugal clutch, which engages the chain when the throttle is squeezed but lets it go when the engine is idling, bringing the chain to a standstill.

In my cabin is a small aluminum wedge that is gouged and scratched. It is screwed into the wall and acts as a hook where I hang duck and goose calls. It's a retired chain catcher from a 372 XP Husqvarna, a souvenir from one of my favorite saws. The job of a chain catcher is to block a derailed or broken chain and keep it from lashing back at the operator. While some are wedge-shaped and aluminum, others are roller pins mounted on the right side of the powerhead near the clutch. A sawyer might end up throwing a chain when cutting at weird angles; the chain rides up and over the groove on the bar. Once loose, the chain whips toward your body so fast you do not have enough time to be terrified. You hear a clatter from the loose chain, and that's it. The rear hand guard of a chainsaw widens on the right side, and that is intended to protect users from a loosened or broken chain.

The safety throttle can be engaged only by pushing a trigger and a button at the same time. On the top of the rear handle is a skinny button that is the lock-out or dead man's switch. That must be depressed for the trigger to rev the engine and put the chain in motion. If the engine could be throttled with just the trigger, a stick could get jammed in there while

Parts of a Chain Saw

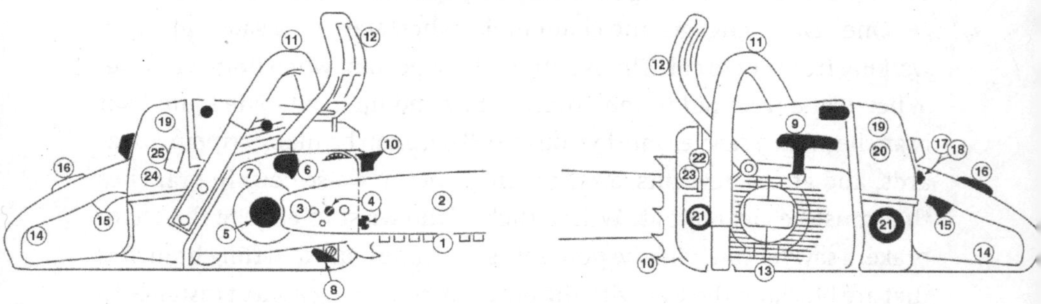

1 Saw chain
2 Guide bar
3 Bar studs
4 Front and side chain tensioners
5 Chain sprocket
6 Chain brake
7 Clutch
8 Chain catcher
9 Starter grip
10 Bumper spikes (dogs)
11 Handlebar
12 Hand guard
13 Gunning marks
14 Throttle handle
15 Throttle trigger
16 Throttle interlock
17 On/off switch
18 Choke
19 Air filter cover
20 Air filter
21 Oil and fuel caps
22 Muffler
23 Spark arrester
24 Spark plug
25 Carburetor adjustments

FROM *CHAIN SAFETY MANUAL*, PERMISSION BY STIHL, INC.

the saw is running and accidentally rev up the engine. Such weird things happen. A friend once told me the story of a duck hunter who set his shotgun on the ground during a hunt and walked off. In the meantime, his dog came back to deliver a retrieved duck. As the dog ran over the shotgun, one of its paws went into the trigger guard, depressed the trigger, and blasted the shotgun, and smoked the hunter in the back of the legs.

The kill switch is another safety feature. It has a firm locking action so it cannot change positions while the saw is running, and it is clearly marked so another person can easily figure out how to turn off the saw in the event of an accident.

On the muffler is a small, blackened screen called a spark arrestor, which stops sparks from spitting out and starting a forest fire. Removing the spark arrestor will increase saw performance—especially if the screen is gummed up; however, a spark arrestor must be in place when using a saw on government land. The US Forest Service has extensive guidelines about this device, and rangers can fine operators of saws without spark arrestors.

One safety feature that gets little recognition is the antivibration system on modern saws. In years past, saws vibrated terribly, and this caused a condition known as white finger, a form of Raynaud's syndrome where a person's hand and fingers turn white due to decreased blood flow. Symptoms include tingling, numbness, color loss during cold weather, and decreased grip strength, dexterity, and feeling. Vibration white finger is now considered an industrial injury. Stihl patented an antivibration system for saws in 1964 that used rubber spacers between the engine and the handle. Other manufacturers followed suit with their own designs, including coiled springs to absorb the shock.

Felling dogs are the most wicked-looking part of a chainsaw. They look like a ninja's throwing star that's been cut in half and bolted onto the front of the powerhead. They are also called bumper spikes, bucking spikes, and dog teeth. Not all saws have them. Many electric chainsaws do not. Some saws have one dog; others have two on each side of the bar. One can add custom felling dogs that are big and aggressive in design, but the bigger spikes reduce the length of the bar that comes into contact with the wood. These metal spikes stabilize the saw during felling and bucking. The spikes dig into the tree bark and act as a pivot point for both horizontal cuts (felling) and vertical cuts (bucking). When the spikes are stabbed into the tree during the felling process, this lessens the weight of the saw and holds the

Felling dogs—the steel spikes attached to this saw's powerhead—help a sawyer stabilize their saw when cutting. Custom felling dogs that are more aggressive than the original manufacturer's parts are popular saw upgrades.

saw steady. "Dog tracks" are left behind as the saw is rotated into the wood and the felling dogs leave small gouges parallel to the kerf.

When a sawyer is bucking, they may tend to use the dogs as a crutch for a dulling chain. The duller the chain gets, the more the sawyer leans into the saw using the dogs as a lever. Although the dogs act as a useful anchored fulcrum, using this leverage and bearing down on the saw will strain and bog down the saw. This is a sign to break out the file for sharpening. A sharp chainsaw will pull itself into the wood, and the sawyer need only guide the bar with a firm hand.

Saw Variants

Polesaws are essentially chainsaws on sticks, meant for cutting out-of-reach limbs. They can be gas-powered, battery-powered, or hydraulic. Hydraulic polesaws are most often found in bucket trucks and used by professional arborists who spend more time working aloft than they do on the ground. They're powerful, with a high torque, and they make a high-pitched whirring sound. The bars on most polesaws are short, and arborists working in buckets often have plastic chainsaw holsters that hook onto the edge of their buckets. When they encounter a thick branch, they tuck away the polesaw and use a chainsaw with a longer bar instead.

Top-handled chainsaws are for single-handed use. On a normal chainsaw, the hand bar and guards are purposely designed and balanced for two-handed operation. It is very awkward, unsafe, and ergonomically difficult to operate a normal chainsaw with one hand. The top-handled versions are popular with tree-climbing arborists who must make cuts while balancing and bracing themselves in challenging positions. The top-handled saw *can* be used with two hands, as there is a wraparound handle on the left side of the powerhead, but two hands are not required. The throttle controls are on the top handle, so the saw can be picked up and used like lifting a lunch box.

Rescue chainsaws are for firefighters and other first responders. The chains on these saws have thick cutters, drive links, and tie straps for cutting through metal sheeting, doors, roofs, and even glass. Some have a depth limiter on the bar, so the operator can make a plunge cut and the bar will penetrate only so far. Husqvarna's line of rescue saws forgoes the bar and chain in favor of a circular saw mounted to the powerhead. Similarly powerful concrete and utility saws are used in the construction and demolition industries. Some of these chains are designed to grind more than cut, and some concrete saws are water cooled in order to combat the heat and the dust. These industrial application saws can also be pneumatic, electrical, or hydraulic. Divers use hydraulic-powered underwater chainsaws for cutting timber on docks, piers, and bridge pilings. The chainsaw revolutionized the timber industry, but like the homesteaders, farmers, and backyard woodcutters, workers in other industries have adopted and tweaked the design for their particular needs.

4

The Sound of Horror

The Saw as a Symbol of Mayhem

The most influential chainsaw movie ever made was inspired by a maniac who used to live about thirty minutes down the road from me. Plainfield, Wisconsin, was the home of Ed Gein, a body-snatching grave robber and murderer. In 1957, Gein was arrested, and law enforcement discovered his farmhouse of horrors. He confessed to killing two women, though he was suspected of killing many more, and he also dug up from the local cemetery the bodies of women who resembled his mother. Gein dismembered the corpses, hoarded the body parts, and fashioned masks, leggings, and other items from human skin.

Not far from Plainfield is the small city of Weyauwega. In 1953, author Robert Bloch and his wife, Marion Ruth Holcombe, moved there and settled down. Bloch was an author of crime, horror, and fantasy stories. His early mentors were H. P. Lovecraft and August Derleth, the literary virtuoso based in Sauk City who founded Arkham House, a press that would later publish Ray Bradbury and Conan the Barbarian creator Robert E. Howard.

By the time he landed in Weyauwega, Bloch had published numerous short stories in pulp magazines and five novels. His next novel, *Pyscho*, was loosely based on Ed Gein, and the Lakeshore Motel in Weyauwega was the inspiration for the Bates Motel that is the setting for the story. The novel was published in 1959 and went on to become his most well-known work. The Lakeshore Motel is long gone, but an Art Deco neon sign for it remains standing in a small meadow visible from US Highway 10. With its rust and missing neon letters, it is an eerie landmark.

The long-gone Lakeshore Motel in Weyauwega, Wisconsin, was the inspiration for the Bates Motel in both the novel and film versions of *Psycho*. That film and the real-life story of Ed Gein influenced and inspired *The Texas Chainsaw Massacre*.

The movie rights to the book were Bloch's first sale to Hollywood. Alfred Hitchcock directed *Psycho*, and it became a classic in cinema history, regularly listed as one of the greatest films of all time. The 1960 film blew the doors open for the nascent slasher film genre, and it influenced horror films in years to follow.

One of these films was *The Texas Chainsaw Massacre*, which debuted in 1974. Ed Gein was an inspiration for the character known as Leatherface, a hulking, mentally disabled killer who wore a mask of human flesh, a butcher's apron, and, oddly enough, a necktie. He was part of a family of inbred cannibals. Leatherface wielded a Poulan 306a chainsaw.

Claude Poulan is known for inventing the bow bar, the curved track that the chain circles around. To picture this, imagine a tennis racquet affixed to the powerhead of a chainsaw in place of the bar. The chain runs through an enclosed top curve that acts as a guide, and when the chain comes around to the bottom of the racquet, the teeth are exposed for cutting wood. Poulan came up with the idea in Texas when he was the boss of a pulpwood crew composed of German prisoners during World War II.

After the war, he founded Poulan Chain Saws in Shreveport, Louisiana. The Poulan 306a was introduced to the public in 1970 and was considered to be a fine lightweight chainsaw in its time. It was discontinued in 1980.

The Texas Chainsaw Massacre pushed chainsaws into the public consciousness. Consider the timeline of chainsaw development. During the World War II era, Andreas Stihl scaled the chainsaw into a one-man device, and his design was widely copied. Right after the war, Joseph Cox invented his "bug chain," which changed the entire industry. Many manufacturers raced to perfect the chainsaw in the 1950s and 1960s, but their saws were still heavy, awkward, and difficult to maintain. It took another decade before the chainsaw was refined enough to be sold to the general public as a user-friendly multipurpose tool. And right at that moment, in the early 1970s, Leatherface emerged as an early adopter of this technology. The 1974 film portrayed the chainsaw as a fearsome tool of immense destructive power.

In a 1986 interview on *Entertainment Tonight*, the critic Joe Bob Briggs called the film the greatest drive-in movie in history. The many sequels and remakes (nine at the time of this book's publication) have added to a sort of chainsaw death cult over the years. In *The Texas Chainsaw Massacre 2*, Dennis Hopper plays a former Texas Ranger and squares off with Leatherface in a chainsaw fight. It was a tool built for bloodshed. It was a tool built for Hollywood. Characters revved chainsaws to show they meant business, much like characters racking rounds into the chambers of pump-action shotguns. These gestures are so widely copied, they're now cheesy gimmicks.

Chevy Chase pokes fun at this cliché as Clark Griswold in *National Lampoon's Christmas Vacation*. His McCulloch 610 makes numerous appearances. To trim an oversized Christmas tree in his driveway, Clark ominously emerges from his garage holding his saw and wearing a hockey goaltender mask (a nod to Jason Voorhees and the *Friday the 13th* franchise). When that tree goes up in flames, he fells a replacement evergreen next to his house, sending it crashing through his neighbor's window. As Clark descends into madness after so many Christmas mishaps, he notices the wobbly cap on the stairway newel post—a reference to George Bailey's loose newel post in *It's a Wonderful Life*. Clark starts his McCulloch with one pull, levels the bar, slices off the cap, kills the engine, and announces,

"Fixed the newel post!" The line is so funny and memorable that one can buy T-shirts, posters, and stickers adorned with a chainsaw and the "Fixed the newel post!" exclamation. In another beloved Christmas film, *Die Hard*, a bad guy uses a chainsaw to cut through conduit holding communication wires in Nakatomi Plaza. It's an Echo QuikVent, a powerful drop-angled rescue saw used by firefighters to ventilate (as the name suggests) burning structures.

One of the most dreadful chainsaw scenes ever filmed was not in a horror movie but in Brian De Palma's 1983 film *Scarface*. In a coke deal turned sideways, Tony Montana (played by Al Pacino) is held at gunpoint while his friend Angel is tied up in the bathroom shower. With a Beretta pistol pressed against his head, Montana is forced to watch his friend get butchered by a Colombian gangster named Hector the Toad. He uses a Homelite XL-2, a lightweight, compact saw that for some reason was painted yellow for the movie. What makes this scene so horrific is that the chainsaw is never shown cutting into human flesh, but the camera work captures the blood spatter, the terror of Angel and Tony, and the psychopathy of Hector the Toad.

The best film that portrays a chainsaw as a tree-cutting tool is the 1971 drama *Sometimes a Great Notion*, based on Ken Kesey's novel of the same name. It is a complex and epic story about a logging clan in Oregon, and the realistic scenes of felling and harvesting trees are my favorite parts of the film. It portrays the physically demanding aspects of the job of a choker setter, the person who attaches cables to logs for skidders or other machines to tow. The family uses McCulloch saws—and not always in the forest. In one scene, Paul Newman cuts the desk of a union boss in half with his long-bar McCulloch Super Pro 125.

Chainsaws have appeared in television shows, videogames, and many B movies, almost always as symbols of power and destruction. Chuck Norris memorably stops a chainsaw with his bare hand in the 1996 film *Forest Warrior*. As a logger cuts into a trunk, Norris pops out from behind the tree and grips the tip of the bar, locking up the chain with superhuman strength.

One of the bestselling manga comics of all time is *Chainsaw Man*. Its creator, Tatsuki Fujimoto, invented the character of Denji, who has chainsaw bars fused to both arms and one sticking out of his forehead. He goes

CHAIN SAW AGE

WORLD WIDE SINCE 1952

DECEMBER, 1971

In This Issue

Zip-Penn Introduces New Laser Direct-Mail Plan

McCulloch Holds Annual Meeting In Las Vegas

Gallery Of Industry Men At New York Hardware Show

Chain Saw Instruction Given Students By Distributor

McCulloch Markets New Big Timber Chain Saw

Stihl Distributor Has Show On Seattle Worlds Fair Site

Chain saws get top billing in advertising, publicity, action and drama — for the first time since their development — when movie star Paul Newman essays role of professional logger in major motion picture, "Sometimes A Great Notion," that premiered in theaters all over the nation in the latter part of December. See story Pages 3, 4, 5 and 8.

The best pop culture portrayal of a chainsaw being used for logging is by Paul Newman in *Sometimes a Great Notion*. He wears a safety helmet in the 1970s film, but eye and ear protection and high-visibility clothing were not commonly used at the time. COURTESY OF *CHAIN SAW AGE* MAGAZINE, HATTON-BROWN PUBLISHERS on various adventures that turn out bloody. It spawned an anime series, a film, a soundtrack, a novel, a stage production, and Chainsaw Man action figures.

In 1983, Music Television (MTV) incorporated the chainsaw in a brief commercial spot to illustrate the difference between what televised music had previously looked like and what the new channel was bringing to American culture. It starts with an image of a black-and-white television, on which a dowdy couple sings a sappy song. Sitting next to the television is a red chainsaw with a bar that glows like a light saber. A young woman in a body suit pull-starts it, picks it up, and cleaves off the screen in a shower of sparks, revealing a glowing MTV logo. In the early eighties, chainsaws were just finding their way into mainstream America. MTV clearly chose the chainsaw as a symbol to show how it was changing the country's cultural landscape by cutting away the old and bringing in the new.

On MTV's sister channel, VH1, the chainsaw made another appearance in the early 2000s during the filming of a reality show called *Surviving Nugent: The Ted Commandments*. The rocker/outdoorsman is a well-known showman, even swinging from vines at some of his concerts. In one episode, Ted Nugent cuts his leg with a saw, requiring forty stitches. In the 1988 film *State Park*, Nugent also appeared in a rocking cameo with singer Rachel Sweet for a performance of "Love Is Like a Chainsaw."

In the world of music, chainsaws are often mentioned in lyrics to make the musician sound tough—even when the skinny-armed singer is not. The rapper Eminem frequently mentions chainsaws in his songs and has been photographed holding a Husqvarna. Yes, he looks violent, but he also looks like a poseur. One can purchase a Marshall Mathers action figure, chainsaw included.

In the most popular chainsaw song of all time, "The Lumberjack" by rock band Jackyl, lead singer Jesse James Dupree revs a gas-powered chainsaw throughout the song and even plays a solo on the saw. Unlike chainsaws in horror films and violent music, this song is full of goofy sexual innuendos, with the lyrics ending, "But I ain't jacked my lumber baby / Since my chain saw you." Dupree is also known for playing "The Star-Spangled Banner" with a chainsaw. He tends to favor Stihl saws for his performances.

With so many forms of pop culture media convincing audiences that chainsaws are tools of mayhem, it makes sense that other entertainers look to chainsaws for their brutish titillation value as well. This is what T. J. Howell did. Out of all of the objects in the world that can be tossed into the air and juggled, chainsaws are the objects he chose. Starting in the 1980s, he joined the ranks of a small group of elite entertainers who practice this highly specialized performance art. Chainsaws are dangerously lopsided objects to juggle, and that is a turnoff for the timid. They are also heavy, and that culls out the weak. In addition, the price of entry is the cost of at least one chainsaw, which is far more expensive than bowling pins or bean bags. The learning curve is not for beginners.

Howell grew up in Leitchfield, Kentucky. His father was a bear trainer, and his mother was a Danish trapeze artist in a circus. He started riding unicycles and juggling when he was twelve years old. By the time he was sixteen, he was getting paid for these skills, and he has been a professional performer ever since. He has juggled while ice skating, while riding a unicycle, and while eating an apple. He's juggled everything from Ping-Pong balls to rings to clubs. He's performed all over the United States and on cruise ships. When he was twenty-one years old, he overheard some guests on a cruise mention that they'd watched a guy on television who juggled chainsaws. Howell had never heard of him. The guests asked if he could do it. He'd never considered such a thing before, but he took it as a challenge.

He found a picture of the man on television who juggled chainsaws and determined that the man had mounted handles onto his saws. Howell cut an aluminum baseball bat and mounted the bat handle to a chainsaw using a bracket. In this first model, he used a McCulloch, and it was way too heavy. He had used chainsaws before and understood the basics of how they operated. He wondered what would happen if he missed a catch.

"At Goodwill, I bought an old mattress," Howell explained in our interview. "I was going to practice flipping [the saw] to see what happens. I gave her a big flip, and it freaked me out. When it's running, the centripetal force of it made it want to do something like a figure eight. Woosh! It's a good thing I had that mattress. I just backed off. At that time, the chain wasn't moving, and it was heavier than all get-out. There were things working against me. I picked it back up and turned the idle down and let this chain run. It helped get rid of that gyro effect, but it was still stupid heavy."

T. J. Howell juggled chainsaws for fifteen summers at Tommy Bartlett's Water Ski & Jumping Boat Thrill Show in the Wisconsin Dells until its closure in 2020.

He went shopping for a smaller chainsaw and bought a McCulloch Mini Mac, a small, chunky yellow saw manufactured in the 1970s. Howell affixed another baseball bat handle to it, but the Mini Mac was still too heavy. He went to a machine shop, had the bar shortened, and had holes drilled into it. That lightened it enough to make a noticeable difference. Howell started to work the chainsaw into his juggling routines. At his shows, he added just enough fuel and bar oil to get through his performance—anything more was just unnecessary weight. He was able to flip the Mini Mac while juggling two apples. Happy with that milestone, he bought two more Mini Macs. The next challenge was how to get all three in the air at the same time. He could only hold two saws at once. But then he discovered a tip guard, a metal semicircle shield that can be installed on the tip of a bar to prevent kickback.

He explained, "I put on that tip guard and leaned [the third saw] against my leg because the tip guard would allow it to still run, and then I would throw up the first one way higher than normal and reach down and grab the one leaning against my leg. Starting and stopping was the biggest challenge of chainsaw juggling. Somewhere along the line, you've got to catch all three. I never got rid of that gyro effect. The thing that worked for me was to let [the chains] run. Don't lock the chain up, that made them weird. It ran the best without the chain locked. If I flipped it and I didn't grab that baseball bat handle and happened to grab or touch the throttle, it would start whistling Dixie."

Howell estimates that over his career, he's broken six chainsaws by underflipping them. "There's no forgiveness," he said. "If I underflipped it, I knew . . . aw shucks! (That wasn't what I said.) I had to pull back, and it hit the ground, and then they explode. It never affected my handles—it was always busting the outer plastic shell of the chainsaw, where you pull-start it. If it happened to land on that, it was doomsday. Finally, after I got tired of buying new parts, I went to a chainsaw repair place in Oregon, Wisconsin, and there was an old guy. I didn't tell him I was a juggler—he'd probably look and laugh at me. I said, 'I have a Mini Mac McCulloch and I need to buy some parts.' And he said, 'I haven't heard of that chainsaw in a long time.' He had two boxes of them. I said, 'How much do you want for all those parts?' He gave them to me. It was the funniest thing. He said, 'I can't believe you even have one,' and then I told him what I did. He ended up being a super nice guy, and I even ended up juggling one for him. I wanted those parts for when I was practicing and I broke the pull start or something and then I had a backup part."

Years later, Howell discovered some Echo chainsaws that were lighter and smaller than the Mini Macs. He bought three of them. Each weighed a little more than five pounds. With these, he also shortened and drilled the bars and added the bat handles. "I can juggle those things like nothing now," he said. "Those Mini Mac McCullochs were very challenging. That's why you don't see many people doing it. And it's expensive. There aren't too many people that are going to throw it up, bust it, and then say they're going to try it again. It's not very common. If you see someone flipping a chainsaw, it's going to be a street performer. Normally, [they] will be [juggling] one. One serves the same purpose of three, but I believe you either go big or stay home. I felt three is the way to go. Whoever tries to do three, they have to figure out what I figured out."

Howell moved to Wisconsin in 1988 and bought a home in Baraboo in 1996. The state has been his home ever since. He performed for fifteen summers at Tommy Bartlett's Water Ski & Jumping Boat Thrill Show in the Wisconsin Dells, the seasonal show started in 1952 that brought water-ski spectacles and stage performances on Lake Delton to millions. "It worked out so nice at the Tommy Bartlett Show," Howell said. "You're outdoors and people are thinking about the woods and looking at pine trees."

The Tommy Bartlett Show was staged twice a day from Memorial Day weekend through Labor Day, rain or shine. The rain was a problem for Howell. He had black tape on the baseball bat handles, and they were "slipperier than snot" when wet. He eventually replaced the tape with tennis racquet wraps for better grip. "I do a lot of different juggling and I'm always focused," he said, "but when it comes to chainsaw juggling, I'm *focusing*. I've nicked myself quite a few times but no stitches—only because I was too confident."

On stage, Howell would start his performance by pull-starting the saws and letting them rumble for a moment. He added extra oil to the gas mix to create more smoke for special effects. To show the audience he was authentic, Howell cut two-by-fours set up on sawhorses. "I had to let everyone know that these are real and I'm cutting through a piece of wood and I'm not faking that," he said. Apparently, the stagehands never liked that because they had to sweep up all the sawdust. In his saw-juggling routine, Howell averaged fifteen catches, which was about a minute of juggling. Over the course of his career, he dropped the saws just a couple of times on stage.

Howell also experimented with battery-powered chainsaws in the late 2010s. He purchased three Makita saws and built handles for them, and they lived up to his expectations. They were light, they cut well, and they were easy to juggle. "The downside was I needed the noise. There was no way of getting around it," he said. "I tried it once, and the producer guy asked, 'What was that?' It looked good. It was dangerous? Yes, it was dangerous. But it just didn't have the wow factor. They needed to bark."

Just as filmmakers use chainsaws as horror props and rappers use chainsaws to look tough, politicians use chainsaws as public relations tools.

Argentina's president, Javier Milei, is a right-wing libertarian—he has called himself an anarcho-capitalist—and a big part of his 2022 campaign platform was about cutting: cutting the central bank, cutting numerous government ministries, cutting public spending, and cutting state-owned companies. That June, he revealed his Chainsaw Plan. At political rallies, he has waved his *motosierra* around with a bar custom-painted with his last name on it. However, some photos captured him wielding a saw with no chain. If the chainsaw was a symbol for his political vision, this would make him literally and figuratively toothless: dull, weak, and ineffective. Yet, the Chainsaw Plan did ultimately result in his winning the presidency in 2023.

American politician Robert F. Kennedy Jr. used a chainsaw to slice the head off a dead whale in the midnineties. His daughter Kathleen "Kick" Kennedy discussed the saw job in a 2012 interview with *Town & Country Magazine*. The dead whale had washed up in Hyannis Port, Massachusetts. After Kennedy cut off its head, he strapped it to the roof of the family's minivan and brought it to his home in New York. Kick explained that it was normal for her father, an environmentalist, to collect and study animal skulls and skeletons.

I've found myself in similar situations. When I lived on the southern coast of South Korea, I once came across a dead minke whale. It was about ten feet long, and it had washed up on a small beach where I was spearfishing, which was off the beaten path. I thought it would be a shame for it to rot, and I pondered how I could preserve its skeleton and reconstruct it, like those found in museums of natural history. But I passed on that project. The only tool I had handy was my dive knife. But if I'd had a chainsaw, I might have tried it. Years later, in the Wisconsin woods, I came across a dead eight-point whitetail buck. I presumed the deer had died from a car-crash injury due to its proximity to a highway. Its carcass was already decomposing, and parts of it had been torn by scavengers. I returned with my chainsaw, cut off its antlers, and later used them to make coat hooks.

President George W. Bush was also known for using a chainsaw to establish some blue-collar credibility. On August 25, 2001, he invited White House reporters to his 1,600-acre ranch in Crawford, Texas, for a tour, and he fired up an Echo CS-3450 to cut up a dead hackberry tree. Bush has also been photographed using a Stihl saw. A reporter from the *Tampa Bay Times* noted that "the only thing that possibly made Bush's chain saw work seem solitary was a pair of safety ear plugs. Standing around and watching were three Secret Service agents, a doctor and nurse, and a military aide who always keeps the nuclear-weapon codes within reach."

Though Bush wore a cowboy hat, blue jeans, leather boots, work gloves, and sunglasses for the photo op, he did not wear any saw chaps. The leader of the free world and the holder of the launch authority for the nuclear football was one preventable chainsaw accident away from a hospital visit. At least he had a doctor and nurse standing by. As for the rest of us, we do not have such luxuries in the woods. There is no audience for us, except perhaps our workmates, and there is real work to be done. If blood is spilled, it is usually the blood of the operator.

5

The 110-Stitch Cut

The Saw as a Lacerater of Human Flesh

In fiction and film, chainsaws are commonly used to hurt other people. But in real life, chainsaw-related violence is extremely rare. In almost all real-life incidents, chainsaws hurt the people who use them. According to the US Consumer Product Safety Commission, the average chainsaw injury requires 110 stitches. A case report in the March 2004 *Journal of Forensic Sciences* noted that the US has an average of 28,000 chainsaw-related injuries each year. A 2017 fact sheet published by the Centers for Disease Control and Prevention put the number at 36,000 injuries.

The majority of these injuries happen at home, which leads me to conclude that many of those injured do not use chainsaws for a living. A 2018 study in the *American Journal of Emergency Medicine* titled "Weekend Woodsmen" compared the patterns of US injuries from chainsaws to those from axes. With data gathered from 18,250 patients, the authors found that 65 percent of saw injuries happened at home and 33 percent were at unspecified locations. Accidents at schools, recreational areas, public property, highways, and farms and ranches made up less than 1 percent of the injuries. As for the unspecified locations, my guess is that the saw injuries occurred on private property away from the home, such as cutting firewood at grandpa's cabin or, for professionals, logging a privately owned tract of land or trimming trees on a backwoods right-of-way.

Among the different types of chainsaw injuries, lacerations were the most common at 68 percent. Fractures, abrasions, contusions, avulsions, amputations, and sprains/strains made up the rest in smaller numbers. The injuries caused by axes were very similar to those caused by saws, both in type and in location where they occurred.

The main culprit in many chainsaw injuries is kickback. The lower tip of the bar, called the attack, is safe and useful and used for boring cuts. The upper tip of the bar is called the kickback, and a sawyer must remain cognizant of this part of the bar at all times. As the chain on the top of the bar moves forward, it rounds the tip and begins coming back at the sawyer. When the top tip of the bar touches an object, instead of cutting, the teeth seize up for a split second and the saw jerks back violently toward the sawyer. It happens faster than a person can react.

The solution to preventing kickback and developing saw safety-mindedness is simple. In fact, it is identical to the firearm safety rules at the Gunsite Academy, the renowned firearms-training school:

1. All guns are always loaded.
2. Never let the muzzle cover anything which you are not willing to destroy.
3. Keep your finger off the trigger until your sights are on the target.
4. Identify your target and what is behind it.

Applying this to chainsaws, the rules would look like this:

1. All saws can always cut you, even when they're not running. A sawyer can trip and fall on the saw. When the saw is sitting on the ground, a person can trip and fall on top of it (laying the saw with the bar flat on the ground helps prevent this).
2. Never let the chain touch anything that you are not willing to destroy.
3. Keep your finger off the throttle when you are not cutting. Keep the finger indexed along the grip area as if the chainsaw were a handgun or rifle. Many inexperienced (and eager) sawyers love to rev the engine and hear that two-stroke sound when the saw isn't touching wood.

4. Identify your target and what is behind it. This rule is exactly the same for guns and chainsaws. Most of the kickbacks I have experienced were caused by an object hidden under or behind the wood that I was cutting.

The most famous case of chainsaw kickback happened in 2014 to twenty-one-year-old James Valentine, a tree climber from Pennsylvania. Valentine was cutting in an awkward position while aloft in a Scotch pine. His saw kicked back at him, and the bar lodged into his left shoulder and neck. He killed the engine and left the saw in place as coworkers helped him to the ground. They removed the powerhead of the saw, but he went to the hospital with the bar and chain stuck inside of him. Amazingly, it missed his carotid artery, spinal cord, trachea, and esophagus, and his shoulder took the brunt of the trauma. Valentine stayed conscious the entire time while doctors removed the bar and chain. He walked away with thirty stitches. His gruesome x-rays were heavily reported in the national and online news.

For decades, the US Consumer Product Safety Commission has produced a chart of the human body showing the most common body parts that get whacked with a chainsaw each year. Baseball teams often analyze spray charts, which are diagrams of every ball put in play throughout the season. The charts can be filtered by ground balls and fly balls and other types of hits. They can reveal players' batting tendencies and help predict the probability of where the next hit ball will end up. In the case of the Consumer Product Safety Commissions' diagram, blood is the thing being sprayed. An outline of a body is marked by small dots indicating saw strikes. The two areas struck by saws most often in 2023 were the hands and fingers (35 percent) and the legs (32 percent). And the majority of the strikes were on the left leg and left hand.

Since chainsaws are designed for right-handed use, it's not surprising that the left leg gets struck quite often. What is odd are the small red dots on the chart that cover the left hand like chickenpox. The saw is a power tool meant for two-handed use, and the left hand should never stray up toward the bar and chain on a running saw.

After hands, fingers, and legs, injuries to the torso and arms were the most common, with about 14 percent of strikes, followed by the head and

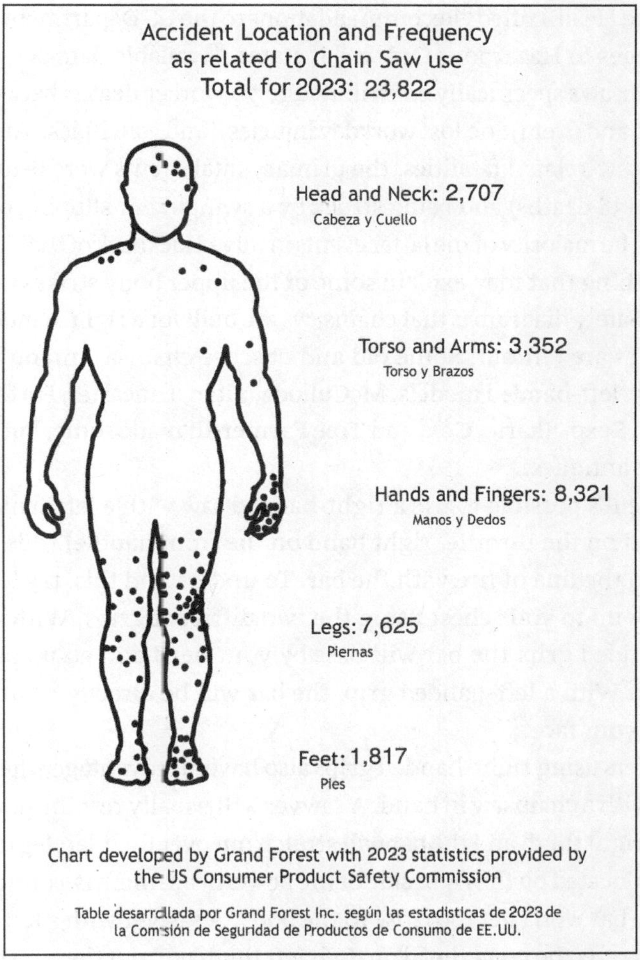

The parts of the body most likely to be cut by a chainsaw are the legs, hands, and head. Saw chaps and cut-resistant gloves can offer protection to the legs and hands, but chainsaw strikes to the head, which are the result of kickback and a failure of the operator, can be prevented only with proper training. COURTESY OF GRAND FOREST, INC.

neck area, with 11 percent of strikes. These are most likely the result of kickback. The foot area was the site of about 8 percent of strikes, and most of those were on the left foot.

The most authoritative statistic about deaths caused by chainsaw accidents is from a 2002 white paper by the National Institute for Occupational

Safety and Health titled "Recommendations to the US Department of Labor for Changes to Hazardous Orders." It states, "Available data demonstrate that chainsaws specifically contributed to 70 worker deaths between 1992 and 1997, and over 1,600 lost workday injuries," and concludes: "Among the 70 chainsaw-related fatalities, the primary fatal events were being struck by a tree (38 deaths) and being struck by a swinging or slipping object (16 deaths). The majority of the latter events involved kickback of the chainsaw."

One thing that may explain some of the upper body strikes on the US Product Safety diagram is that chainsaws are built for a right-handed world and lefties are left out. Some old and obscure chainsaw manufacturers built a few left-handed models. McCulloch, Titan, Lancaster, Porter-Cable, Le Sabre, Rexo, Skarie, Cox, and Tree Farmer all made some, but they are now rare antiques.

While it's possible to use a right-handed saw with a left-handed grip (left hand on the throttle, right hand on the front handle), this puts the sawyer in the line of fire with the bar. To understand this, try hoisting a chainsaw up to your chest using the two different grips. With a proper right-handed grip, the bar will pass by your head and go up over your shoulder. With a left-handed grip, the bar will be directly positioned to chew up your face.

Sawyers using right-handed grips also have an advantage when taking a break with a chainsaw in hand. A sawyer will usually rest the powerhead on their right thigh and the bar will stretch out over their left leg, but since the bar is located on the right side of the powerhead, there is a protective air gap—the bar won't get too close to the body. A sawyer with a left-handed grip who rests the powerhead on their left thigh will put the bar very close to their right leg. Another safety concern for left-handers is that the kill switch can be instantly flicked off with a user's right thumb. If held left-handed, one must let go of the saw with one hand to hit the kill switch.

Of course, there are rare times when even I, a righty, have changed hand positions on my saw, usually when I'm flushing stumps close to the ground. When I am kneeling and my body is contorted into an awkward position because of nearby rocks, trees, or other annoyances, I will switch grips to make the work easier under those conditions.

The one advantage a lefty does have with right-handed, gas-powered saws is starting them with the pull cord. These are almost always on the left side of the powerhead. Lefties can yank the starter cord with their strong hand, which is useful because one of the most exhausting and frustrating exercises in the world is yanking the pull cord of a chainsaw (or any machine) that does not want to start. It can make a person go insane.

The proper way to start a gas-powered chainsaw is with the chain brake on. The safest way is to press the powerhead on the ground with one hand and yank the pull cord with the other. Sometimes a sawyer can tuck the tip of a work boot, if it will fit, onto the rear hand guard to help stabilize the saw. Another method is the crotch start, where a sawyer tucks the handle end of the saw between their legs and squeezes it with their thighs to stabilize the saw and then pulls the cord from that position.

The most frowned-upon method of starting a chainsaw is the drop start or throw start. Here, the sawyer holds the saw out and away from their body and, in one quick motion, pushes the saw downward while pulling the starter cord upward. This method is considered unsafe because the saw is not stabilized and the bar jerks erratically. If the chain brake is not on, the chain will spin, and the sawyer will not be in control of the saw with two hands.

Some prize the flair of drop starting over safety, however, like Wayne King, a reporter who covered a chainsaw competition in 1979 for *The New York Times*. He wrote, "The weekender lays the saw on the ground, puts a foot on it as if it might snarl and bite, then makes a pull or two on the starting cord, tentative, like a little old lady stroking a Doberman Pinscher. By contrast, Dave Geer, a Connecticut tree farmer who fells trees the way a bear eats honey—by instinct—snatches up a yard-long Homelite 750 and rips it into life with a single fluid motion: Down goes the saw, up goes the cord, and the engine sings like a hornet." Though professional arborist and logging operations usually have jobsite rules banning crew members from using this starting method, many sawyers do it this way and shrug off the safety concerns. It is very common to see, and I have done it myself.

When Trees Strike Back

"Lynchings, casual violence, death in icy rivers or from failure to dodge a widow-maker in the big woods—all these were part of the roaring century when Wisconsin pine was cut," wrote Robert Wells in a 1976 *Milwaukee*

Field projects coordinator Adam Gundlach sets out cones below a widow-maker along a path in the UW Lakeshore Nature Preserve. PHOTO BY BRYN SCRIVER, UW LAKESHORE NATURE PRESERVE

Journal essay about the early days of logging. Nowadays, getting lynched or attacked is unlikely. Taking a dive in a cold river could be possible, but the biggest risk while using a chainsaw is the possibility of getting killed by a tree in motion. A widow-maker is a broken-off tree branch that is suspended in midair, tangled in the branches of the tree it came from or a nearby tree. It hangs there like a booby trap. The danger is compounded by the fact that hard, heavy wood starts moving slowly, then moves very fast. A person could be killed by a tree falling, swinging in an unexpected direction, or bouncing off the ground.

"Let's imagine that life is like a board game," wrote Laurence Gonzales in *Deep Survival*, a comprehensive book on the science of survival. "Some people like to play checkers, some people like to play chess. And those of us who go into the wilderness and engage in dangerous sports are playing chess with Mother Nature." For loggers, arborists, and line-clearance tree trimmers, the board game is like three-dimensional chess. There are many moving pieces, powerful tools are in motion, and there are always unseen threats. Some of the case studies in *Deep Survival* feature people on a one-time outdoor expedition, yet loggers and tree workers face lethal dangers every workday for years. "Survival is a continuous spiritual and physical act that spans a lifetime," Gonzales wrote. Despite the great technological advances that have occurred between crosscut saws and the modern chainsaws of today, one thing has never changed: woodsmen are still getting killed by falling trees. Studying chainsaw-related injuries leads me to make three conclusions: chainsaws injure people more than kill them, kickback is a leading cause of those injuries, and what really kills people are the trees they are working on.

It was fifteen degrees below zero at midmorning on February 15, 2007, when Gary Edinger went into the woods in Emery, Wisconsin, to cut some timber. He was an independent logger in his late sixties who cut trees with a chainsaw. That day, he was working on a cojoined soft maple with full-size stems. He notched one stem and then notched the other. His idea was that one would fall onto the other. Instead, the second stem barber-chaired and fell onto the first stem, which fell but got tangled in another tree. The second stem skidded down the hung-up stem as Edinger scrambled over the stump and out of the way. As it crashed to the ground, it severed his lower left leg, with the stump acting as a butcher's block. He used his belt as a tourniquet to stop the flow of blood, but the belt broke. Edinger expected he would bleed to death, but he did not give up. He crawled to his skidder, bringing his bloody foot with him. He drove the skidder to his truck, got in, and used his right foot to push in the clutch to use the stick shift. He called 911, and the dispatcher sent a sheriff's deputy and ambulance, which intercepted him on a county road. He was later flown by helicopter to a hospital in Marshfield. He now walks with a carbon-fiber prosthesis. He later returned to the accident scene, harvested the tree that claimed his left leg, and wrote of his tale in his book, *The Will to Live: A Saga of Survival*.

On January 2, 2024, Adam Wehling of Mondovi, Wisconsin, was running his Stihl saw on a forested hill in the town of Drammen. The tree he was working on snapped at a point six feet above the ground. The upper half of the cracked trunk hinged over, with some splintered wood fibers still connecting the standing trunk to the fallen upper half that leaned on the ground. He wanted the tree to come down completely, not just so he could cut it up for firewood but also because he figured his children might want to climb on the snapped tree and it did not look safe.

He notched the trunk and made the back cut. The tree turned out to be hollow, and the giant mess fell on top of him—the standing spar that he notched and cut fell onto his right side, and the broken part of the trunk came down on his left side. His body was covered with splintered wood where the tree had originally fractured, and the weight of both separate logs pinned his legs, pelvis, and abdominal area. He pulled out his phone and called 911. He later said his biggest fear was that he would not be found. An Eau Claire County sheriff's deputy was first to arrive, and he ran up the hill toward Wehling. The deputy fired up Wehling's saw and cut the log pinning Wehling's left side, but it was still too heavy to move. When a first responder arrived, the two of them were able to heft the log enough for Wehling to roll out. Incredibly, he walked away with just a few bruises.

Trees in motion almost killed Edinger and Wehling, and their accidents have several elements in common. Both incidents occurred during the felling process. Both men were working alone. And in both cases, their cell phones saved them. But these stories may have ended differently if Edinger and Wehling had used different felling techniques and had not been alone in the woods.

For any type of outdoor activity, the buddy system is wise. It's two people looking out for each other. It's someone having your back. It's the human equivalent of built-in redundancy: two is one, one is none. While I worked as a line-clearance tree trimmer, my company's policy was the buddy system. If one guy on a two-man crew called in hungover, the other man could not work until the general foreman rounded up a guy from another crew to work with him. The felling policy was that all trees required two people: one on the chainsaw and one pulling down the tree with a rope at a safe distance.

The buddy system makes sense for commercial operations, but for a homesteader, solo firewood cutter, or independent logger, it raises the question: does a sawyer need a buddy tagging along every time they fire up a chainsaw? This would certainly be impractical. Having a buddy on an ice-fishing trip or a duck hunt is fun and makes for a safer trip. Getting constantly asked to help a friend logging or cutting firewood would get old very fast. Cutting and harvesting wood in the forest with friends and family can be a rewarding bonding experience, but it is still a hell of a lot of work, and that kind of generosity of time and labor should be requested in small doses. The surrogate for a buddy is a cell phone or walkie-talkie.

The only problem is that remote locations often have no reception, and even if a connection is made, help could be a long time coming.

Based on the OSHA accident archives showing that many chainsaw-related deaths are caused by trees in motion, the buddy system would be most prudent during tree felling. While one buddy uses the chainsaw, the other buddy acts as a grounds person and uses a rope to help guide the tree's fall in the desired direction. To make efficient use of this time with another person acting as emergency backup, a sawyer can target and fell multiple trees during a work session with the goal of simply getting the trees safely on the ground. Once the trees are on the ground, they are no longer in motion and not nearly so dangerous. The dynamic energy is gone, and the powerful mass of the tree rests quietly on the earth. Of course, there are times when a fallen tree can shift or roll, but the biggest danger is eliminated when the tree is horizontal. The next step of limbing and bucking the tree can be done alone if a buddy is not available. You are not completely out of the woods yet, because an accident with the chainsaw can always occur. But statistically, you are much more likely to get lacerated than killed, so you have that going for you. May the forest be with you.

Armor and Aid

Chaps

Ted Nugent could have saved himself some stitches if he had been wearing chaps. Chaps are the most important piece of all chainsaw personal protection equipment. People may shrug them off or even deride them, but when an accident or close call happens, they become lifelong believers. Considering the number of saw strikes to the legs that appear on the US Product Safety Commission's annual chainsaw blood-spray chart, wearing chaps is a very wise decision.

I have cut my leg only once, and I was wearing chaps when it occurred. My crew and I were cutting out a swamp of tag alders in a protected wetland area where heavy machinery was not allowed. Everything had to be cut by hand with chainsaws. Thousands of tag alders needed to be cut down on that winter day. At least the water was frozen, so we could walk on ice rather than be balls-deep in cold mud water. The work was physically demanding; as we worked, we all shed layers until we were in our undershirts. I took a breather, panting in the frosty air. My arms were like rubber, and I leaned the powerhead of the saw against my right thigh for a moment's rest. That's when I felt a hard tug on my left leg. I had let my arms sag in a tired daze, and the bar had turned just enough for the teeth to nick the fabric of the chaps. It ripped the outer fabric, and the Kevlar stuffing snarled into the teeth, stopping the revolving chain. The chaps worked exactly as intended. The foreman informed me that, according to a safety rule, the chaps could no longer be used and had to be thrown out. I volunteered to dispose of the chaps. The foreman smiled and said that was fine as long as he never saw them again. I brought them home and patched the rip with Gorilla Glue. I still wear the chaps to this day.

Chainsaw chaps are affordable and easy enough to put on, but they are cumbersome and bulky. One problem is that they tend to sag after some activity, so in addition to tightening the belt and leg straps, I make it a habit to wear a cased Leatherman tool at the small of my back on my regular belt. I hook the belt of the chaps over the Leatherman tool, and that keeps the chaps from sagging. Chaps are not good for all-day wear. They provide high protection when cutting but high drag for everything else. It's best to wear them only when cutting and take them off for greater mobility when picking up firewood, stacking brush, and doing other physical chores. In winter, I always bring my chaps inside to dry out overnight. Nothing is worse on a cold morning than putting on frozen stiff chaps that are crusted with frost. The same goes for chaps that get soaked in the rain. But I've found that after enough exposure to bar oil and forest grime, the front of the chaps tends to develop a water-resistant patina.

If you want a chaps upgrade, you can buy saw pants with protective Kevlar stitched in. They are more expensive, but those who wear them say they are worth every penny. These are often worn by tree climbers who need the agility that bulky chaps cannot provide. Saw pants are also useful for those who must pick up, use, and put down a chainsaw countless times throughout the workday. It's one less personal protection item to think about, and saw pants save them from constantly taking off and putting on chaps. One fellow I worked with swore his saw pants made the best winter pants. The protective Kevlar padding over the knees, thighs, and shins allowed him to kneel comfortably during ice fishing and making sets while trapping.

Brain Buckets

Falling objects are a huge hazard for those who regularly work with chainsaws. When doing tree work, you could take a branch to the dome at almost any time. One windy day, I was walking through my woods, and I followed a trail under some large oaks that had died from oak wilt. When I circled back down the trail minutes later, a branch as thick as my leg was stabbed into the ground right where I had just walked. It would have split my head open like a coconut. I felled those trees later (while wearing a hard hat) and turned them into firewood.

Some chainsaw-using groups are doctrinarian about wearing hardhats on a jobsite, requiring that they be kept on even during the slightest break on the hottest day. This extreme position is held not for reasons of employee safety, but out of fear of worker's comp claims, lawsuits, and slaps on the wrist from OSHA. James Fallows of *The Atlantic* has written extensively on post-9/11 "security theater," the idea that certain security measures—like the demeaning checkpoints at airports—are mostly empty gestures that give the illusion of security. In my experience, many workplace rules are nothing more than "safety theater" that totally disregards common sense and makes employees feel like children.

Yes, logging and arboriculture are very dangerous occupations. But there can be so many safety rules that it is difficult for those with their boots on the ground to get their jobs done. My advice for people using chainsaws on their own is this: act like an adult, use your own judgment, and if you make a dumb mistake, take responsibility for it and try to learn from it. If you are cutting firewood in an open wood yard where nothing will fall from the sky other than rain, snow, or bird shit, you might decide that a hard hat is not necessary.

Carrying a Saw

As discussed earlier in this chapter, even the simple act of walking with a chainsaw can pose a threat to users. The safest way to walk while holding a chainsaw without a scabbard is to hold the saw by the top handle with the bar pointing behind you. It is simple and easy. Your arm gets tired after hiking long distances, but then you just switch arms.

My favorite way to carry a saw is to hold the powerhead against my hip with the bar angled down and behind me. This style of carry is shown in the woodcut on the cover of Jeff Jepson's excellent book *Groundie: The Indispensable and Diverse Role of the Arboricultural Ground Worker*. Carrying a saw in this manner makes some safety puritans stomp their feet, but I've never heard a good explanation as to why it is unsafe.

You may carry a saw by putting the bar on your shoulder with the powerhead balanced behind you, but this is dangerous. The teeth come very close to the carotid artery, which means the sawyer is one stumble away from getting bloody. Yet many stubborn loggers do this, and there are even leather shoulder pads made and sold for this purpose. The pad even has two slits for suspenders.

The best way to carry a chainsaw for long distances is to use a specialized chainsaw backpack or to modify a backpack for that purpose. Your hands are free, the weight is carried on the back and hips, and your arms will be fresh at the work destination.

Safety Specs

Wood chips fly everywhere during chainsaw work, and even though I am religious about wearing eye protection, tiny particles have still found their way into my eyes. Whenever a tiny bit of debris finds its way over, under, or around my wraparound glasses, I stop work immediately and fetch a saline rinse from the first-aid kit. Though safety glasses with a tight seal do a very good job protecting the eyes, they constantly fog up on hot and humid days. Some sawyers cheat by sliding the glasses down the bridge of their nose like an old-timer wearing reading glasses. Then the sawyer can see, but the eyes are more exposed. The only way to deal with this annoyance, other than stopping every few minutes to wipe your glasses, is to use the antifog chemical that downhill skiers and scuba divers use on their face masks.

Regular sunglasses are not safe enough. Safety glasses should have a Z-87 shatterproof rating, and I recommend having three pairs on rotation: dark shades for sunny days, clear lenses for early mornings and doing work in shadowy areas, and yellow lenses for overcast skies or semishaded spots.

Ear Protection

Chainsaws are as loud as the Motörhead concert I attended. The Department of Defense's Center for Hearing Excellence measures rock concerts at 120 decibels and chainsaws at around 115 to 120 decibels. Hot, or high-powered, saws are even louder. I love the sound of heavy metal blasting off speaker stacks, and I also appreciate the sound of two-stroke chainsaws—just not enough to jeopardize my hearing. Like any loud power machine, a chainsaw requires ear protection. I rotate through three different types, depending on my task. For groundwork, I wear either spongy foam ear plugs or ear muffs, which I also use for shooting firearms. For felling and any job that might involve a falling object, I use either foam ear plugs with a helmet or a special hard hat with fold-out earmuffs mounted on the sides.

Gloves

Be leery of advice from a chainsaw operator missing a finger. Cut-resistant gloves should have an ANSI Level 4 rating or above. The American National Standards Institute (ANSI) ranks safety gloves into nine levels of cutting protection. Gloves in levels 1 through 3 are for light protection, and level 4 gloves are considered appropriate for medium hazards found in many workplaces. For three-season wear, many gloves are of the stretch-fit variety with the rubber-dipped palms. They are ugly as sin, but they work and can be tossed in a washing machine. Cut-resistant winter gloves are few and far between. The best ones I've found are the Superior Glove company's Endura line of goatskin gloves lined with Kevlar and Thinsulate. These gloves will not allow you to stop a spinning chain like Chuck Norris—no glove is truly cut-proof unless you are wearing chainmail—but the Enduras offer enough protection to reduce mangling.

Boots

Both my Redwing and Thorogood boots have chainsaw nicks in the toe area. In both instances, the damage came from cutting claustrophobically thick brush on a steep incline while working too fast. Without the armored toe protection, I might have ended up with bloody socks. Steel-toe or composite-toe work boots are a wise choice when doing chainsaw work. The only trade-off is that the steel toe gets cold in winter and makes your toes numb.

The linemen/logger-style boots are popular, and the high heel can be dug into the earth for traction and stability on uneven terrain. The high heel also serves a purpose similar to the heels on cowboy boots. Like a rider hooking a boot into stirrups, a line worker or logger can lock their boots into climbing gaffs. However, many tree-climbing arborists prefer a work boot/hiking boot/sneaker hybrid.

The best method for keeping leather boots waterproofed and conditioned garners much discussion around a tailgate or campfire. Neatsfoot oil and saddle soap work to an extent. Most modern mink oils no longer contain real mink oil and are falsely advertised synthetic concoctions that are not as good as the natural version. Beeswax-based creams work very well on leather, but they prevent the leather from breathing, and this causes feet to overheat and sweat. My current choice of leather protector is Nor-V-Gen Shoe Oil made in Downing, Wisconsin.

Scabbards

The chainsaw scabbard is another protective item, not just for yourself but for everything that can come into contact with the tool's sharp teeth. The scabbard can protect you if you happen to trip and fall while carrying the saw. The best use for the scabbard is in storage and transport. On one job, I witnessed an unsheathed saw regularly being jammed into the bed of a work truck at the end of each day. The truck was loaded with arborist gear, and it took a long time before anyone noticed the pool of bar oil congealed on the bed of the truck. The chainsaw teeth had gouged the nearby plastic jug just enough to make a tiny leak.

In addition to preventing small cuts, scabbards protect inanimate objects. An uncovered bar and chain bouncing in the back of a truck will snag fabric, gouge plastic, and scratch surfaces.

Nearly all scabbards are made of molded plastic and come with a newly purchased saw. They get cracked, grubby, and nicked over time. They are ugly and utilitarian, but they work and are cheap to replace. I've seen wooden scabbards custom-made for longer bars. The most handsome saw scabbards I've ever seen are handmade leather ones from Traverse Creek in Pennsylvania that have a side pocket with a thong snap for securing a bar tool.

Hi-vis Gear

Wearing high-visibility shirts, vests, or jackets is smart when working with or near others using chainsaws. If you are alone in the woods, your visibility probably doesn't matter except in the case of an accident. The search-and-rescue team might find your bloody body more easily if you are wearing hi-vis clothing. That was Adam Wehling's fear when he was pinned under the log: that he would not be found. The reason early lumberjacks and deer hunters wore red buffalo plaid wool jackets was so they could see each other better. In the September 1894 issue of *Munsey's Magazine*, one lumberjack explained: "Falling trees and rolling logs have caused a long list of deaths; and it is on this account that the woodsman's outer garments are of the brightest colors, blue, green, red, and yellow being the more prominent. The men are thereby able to see one another more distinctly through the thick underbrush, and by a timely warning to avert a great many dangers."

Blood Stoppers

Over the years, I have experimented with countless individual first-aid kits (IFAKs), which are meant to be carried on your person. I've reduced my IFAK down to only one item: the Israeli bandage. It was invented by the military medic Bernard Bar-Natan, and American soldiers discovered its usefulness while serving in Operation Enduring Freedom and Operation Iraqi Freedom. It is an elasticized bandage with a sewn-in nonadhesive pad, built-in pressure bar, and closure clip. The two main dangers while doing chainsaw work in the woods are blood loss and blunt-force trauma from moving trees. There is not much I can do if a tree crashes down on me or a broken-off branch whacks me in the head. Nothing in a pocket-sized first-aid kit will help me in those situations. The only thing I can do is try to prevent those things with prudence, caution, education, and experience. Blood loss, however, can often be controlled with a shrink-wrapped item that fits in your pocket. I treat my Israeli bandage just like my chaps, safety glasses, and ear protection: it's always on me while I'm in the woods. Tweezers, eye drops, alcohol swabs, and all the other first-aid crap can remain in the large kit on the ATV or work truck, which might be a distance from the felling or cutting site. But this one bandage goes where I go. If I cut myself, I want to stop the blood loss before taking a single step.

Other Dangers in the Woods

Steep Slopes: One late-winter day in Green Bay, our crew was tasked with clearing a steep slope covered with snow and ice on a right-of-way. The trick, we discovered, was to cut the brush just high enough to leave small nubs that we could step on like mountain goats, steadying ourselves for the next move upward. After a morning of cutting, the hill was essentially covered with small punji sticks, or booby-trap stakes. I fell down the slope that day and took a hard gouge into the ribs. Even when ice is not a factor, working with chainsaws on inclines is exhausting. The muscles in the bottoms of your feet get tired as they are constantly clenching for balance.

Stingers: I once found myself working along the Milwaukee River near the lost-in-time community of Young America, Wisconsin. It was quaint and out of the way, and it reminded me of Spectre, the secret village in the film *Big Fish*. I flushed a rotted spar that was home to a beehive. I was stung five times before I noticed the gathering swarm. I dropped the saw and ran away. Thankfully, I am not allergic to bee stings, but it was enough venom to get me high on histamine. The saw idled for fifteen minutes until the bees settled down. Then I retrieved it by connecting pole extensions and hooking it with a polesaw head. Now, whenever I'm inspecting a tree for height and hazards, I look for wasp nests in the canopy, and I check decayed wood with crevices for bees.

Ticks: I've been bitten by ticks so many times while working in the woods that I feel as if I'm playing a lifelong game of Russian roulette with Lyme disease. Other than bug spray, there is not much you can do to prevent them from crawling onto you. The only recourse is to remove them as fast as you find them and swab any bite areas with alcohol. Once, while trimming trees near Cauldron Falls, our crew unknowingly got into an area absolutely full of ticks. My two coworkers and I spent the afternoon pulling ticks off each other like lice-picking monkeys.

Urushiol: When cutting with a chainsaw, a fine mist of bar oil sprays off the bar, and sometimes other liquids can, too, such as water from melted snow and tree sap. When a sawyer gets into poison ivy and its cousins, poison oak and poison sumac, however, the tiny flecks spraying onto their skin are caustic toxins. In order to avoid these plants, you must be able to identify them, and it is surprising how many people cannot. Once identified, you can avoid them. But when I encounter poison ivy, I often find it surrounding the tree that I planned to cut, and that makes it pretty much unavoidable. I was immune to urushiol (the oily compound of poison ivy that bonds to the skin) for most of my life, but one day in middle age, my invincibility disappeared, and now poison ivy blisters have become a regular part of my life.

I've found that it's best to just wade in and get the cutting work done because time can be on your side if you move quickly. Urushiol takes six to eight hours to completely bond to the skin, which means there is time to wash it off, and the sooner the better. The protocol I've adopted is to have a water jug near my worksite along with a bottle of dish detergent and a clean shirt. As soon as my work is done in the poison ivy–infested area, I strip the clothes from my upper body and lather up, rinse, and repeat as many times as I feel necessary. Urushiol can also get on equipment, helmets, and chaps; I wipe down those items with a soapy rag and toss my exposed work clothes in the washing machine as soon as possible.

Thorns: Buckthorn, as the name implies, is a shrub with many thorns along its branches. Once, I got a thorn stuck in the meat of my hand where the thumb meets the palm. I removed the thorn and tried to forget about it, but for the next three weeks, my right hand had a mild ache and my grip was stiff and swollen. It was a case of plant-thorn arthritis. I had left behind a small piece of residual thorn under the skin, and this caused a localized inflammation in the joint. The stiffness eventually went away on its own.

Thorny trees and shrubs will always be difficult to work with, and I've found wearing extra protective clothing for thorns is not at all worth the effort. Between chaps, safety glasses, and leather gloves, my body is already mostly covered. Occasionally, a woods worker might get caught in a spot thick with nasty thorns, but that is an exception. The vast majority of thorn-proof clothing is marketed to outdoorspeople, often upland bird hunters, who are delusional and/or overly imaginative when it comes to the vegetative threats they might face. "Brush-buster" pants and "briar-proof" fabrics are usually made of heavy cotton, and I'll tell you a dirty secret about waxed cotton, tin cloth, and duck-canvas clothing: they are insufferable to wear while doing hard, physical work. The material does not breathe well, and the result is a suffocating, overheated feeling. Feeling miserable in order to be covered in stuffy armor is just not worth the trade-off. I do have a waxed cotton bird-hunting jacket, but I wear it as a windbreaker shell on very cold winter days—not for protection against thorns.

Booby Traps: The concept of the line of fire should be ingrained in any woods worker. You can be in the line of fire when a climber above you is sending falling limbs into your zone. You can be in the line of fire when your fellow crew member swings his saw low to the ground as you walk by. Your head and upper body can be in the line of fire when a saw kicks back. With a little common sense, you can simply get out of the way. The problem is that there are so many ways a sawyer can be put in the line of fire without realizing it. You must always keep in mind that the human body can be a target for random—but sometimes predictable—forces.

Another important concept to understand is stored energy. Take, for example, a spring pole. When a large tree falls and lands on a smaller tree, pinning the top of that tree to the ground and bending its trunk in an arch, the bent tree becomes a spring pole. When its trunk is cut, it will snap back. I once accidentally cut a spring pole camouflaged in a tangle of brush and canopy. It released its stored energy and sprang upward, and I was in the line of fire. It knocked off my helmet and gave me a bloody lip.

The proper way to cut a spring pole is to make small relief cuts at the deepest bend in the trunk—the maximum point of tension. Stand out of the predicted line of fire and make small cuts on the bottom compression wood of the spring pole (while taking care not to get the bar pinched) and small cuts on the top tension wood. Eventually, the small cuts allow the stored energy to dissipate, and the final cut can be safely made. Another option is to leave the spring pole alone and work around it. Eventually, the tree pinning it will be cut down to size and can be rolled off the spring pole, which sometimes will spring up like a trebuchet hurling a rock. Spring poles that have been pinned for a long time are harder to predict. Some will not spring up at all if the wood fibers of the tree have become permanently distorted in that bowed-over position. If green suckers are shooting up from the bend of the spring pole, then you can assume the tree has been stuck in that position for at least one growing season.

I found myself in the line of fire once when my crew and I were recovering a skid steer stuck in a ditch of mud and snow. We hitched an old rigging rope to the machine to pull it out. A proper tow strap would have been better, but we didn't have one long enough and we only had to tug the skid steer a few feet, so why not? We hitched the other end of the rope to a Kershaw Skytrim, a powerful machine with tracks like a tank. As the Skytrim pulled the skid steer out of the ditch, the rope quivered with extreme tension. Everyone backed far away. The rope snapped, and it came in my direction like a bullwhip. Luckily, I was at a safe distance, but had I been closer I could have been partially decapitated or at least disfigured.

6

Fast Cuts at Symco

The Saw as a Hot Rod

In 1967, a group of farmers gathered in a central Wisconsin field to witness a demonstration of an old threshing machine and a Rumely Oil Pull 25-40 tractor. They enjoyed it so much, they formed the Union Thresherman's Club and did it again the next year and the year after that. Now this annual three-day event—called the Union Thresheree—has grown so large, the parking area is the size of a few football fields, and there is a camping area for recreational vehicles. This happens every summer in the small burg of Symco in the town of Union. Over the years, the club has continued to create the fictional, retro town of Unionville on private land for the event, and it has evolved into a community that seems to be frozen in time during pre-electrical America. The town runs on water, kerosene, steam, horsepower, manpower, diesel, coal, wood, and hundreds of volunteers. There is a blacksmith shop, saloon, print shop, general store, school, church, post office, cobbler, police station, barber shop, water-powered feed mill, and sawmill. I headed straight for the sawmill when I visited on a hot summer day in July 2023.

Inside the mill, three sweaty men were making wood shingles on an antique machine. I watched them for a while, then circled around to another part of the mill where I found dozens of vintage chainsaws hanging from the rafters like weird chandeliers. There was a Wright reciprocating saw, a David Bradley saw, a Clinton, a Disston, a Homelite, a few two-man saws (like those invented by Andreas Stihl around 1920), a bow-bar saw, and many others. Some were so high up and covered with dust

These antique saws have been saved but not restored. Still, saving them is important, as thousands if not millions of old chainsaws have been scrapped as junk metal over the decades. Old chainsaws like these are becoming rare and valuable finds.

and pigeon shit that I could not recognize their make or model. Attached to one saw was a nest of paper wasps the size of a beach ball. For a moment, the grubby condition of these antique saws annoyed me, but then I reconsidered. Someone had taken the time to recover and save these saws, to hang them in the rafters where they were out of the way, protected from theft, and dry enough to avoid corrosion. Also, one wall of the sawmill was covered with rare old chainsaw advertisements. I felt a moment of nostalgia, remembering walking into a saw shop with my father when I was a boy and smelling cigarette smoke, gasoline, and oily machine parts. Those were the kind of chainsaw ads that hung on the walls of that shop.

I hiked over to a bandstand stage where the antique chainsaw contest was getting set up. In the world of chainsaw racing, there are two schools. In West Coast racing, sawyers make one cut on a very large log using a saw with a large bar. In East Coast racing, sawyers make three precision cuts on a smaller log, and the saw's bar size is negligible—it just needs to be big enough to get the job done. Bigger bars have longer chains, which require more sharpening and have a greater chance of derailing. There is no governing body for this sport. Each competition has its own rules regarding stock, modified, and custom saws, as well as its own safety precautions.

The chainsaw races that get the most attention are the Stihl Timbersports competitions, which are held in different locations throughout the year, and the Lumberjack World Championships held annually in Hayward, Wisconsin. However, the chainsaw race is only part of these events. In the Stihl series, there are six disciplines: the springboard chop, underhand chop, and standing block chop, which are done with an axe; the single buck, done with a crosscut saw; and the hot saw, done with a chainsaw. The Lumberjack World Championships includes those six events, as well as log rolling, boom running, speed pole climbing, axe throwing, and double-buck sawing with a crosscut saw.

The competition I watched at the Union Thresheree featured a mix of old and new saws arranged into classes by engine displacement: 50 cc and under, 60 cc and under, 70 cc and under, 80 cc and under, and 100 cc and under. A special hot saw class existed for the over-100 cc saws. To close the competition, an event challenged sawyers to make the best cut with the exact same saw—a competition not about the power of the saw but about its handler.

Eleven sawyers competed, carrying fleets of McCullochs, Stihls, Husqvarnas, Jonsereds, Dolmars, Homelites, and others with them. The contest is simple: make one downward cut at the end of a log and cut off a cookie, or a round slice of the log, then make an upward cut with the top of the bar and slice off another cookie. The final cut is another downward cut to create a third cookie. Competitors make those three cuts as fast as they can.

The event is timed by three judges with stopwatches. The clock starts as soon as the chain touches the wood and stops as the last cookie drops. On this day, the pine logs were rough-cut eight-inch beams ratchet-strapped to steel sawhorse-like bases. As contestants continued shaving cookies off the ends during each round, the judges released the ratchet straps and scooted the logs forward. They did this until the logs were cut down to blocky nubs.

The saw handling in these events must be precise, tight, and confident. The most common operator error is a wobbling bar on the second and third cuts. Upon dropping the first cookie, a sawyer might bring the bar back up too close to the freshly cut end, causing the chain to skid off the wood and losing a precious second getting the teeth back on the wood.

Each saw is different, not just in make and model but in performance and personality. Some saws at the thresheree were so old they did not have chain brakes. Some were smoky, filling the air with blue exhaust. All of them sounded different, from a farty rumble to an eardrum-piercing shrill. The bars were of various sizes, and I noticed that length seemed to make no difference in cutting speed. Some of the saws got noticeably bogged down during the cutting. One sawyer walked away after his first cut, as his saw couldn't muster the power. Other saws zipped right through. Most impressive was the spray of wood chips thrown by the saws, caused by a combination of teeth sharpness, raker depth, and engine power. Get those three factors in the right balance and the sharp, speedy saws sprayed chips like a rooster tail from a speedboat. In contrast, the not-so-sharp, anemic saws sputtered them out like a low-pressure garden hose.

On this particular day at the Union Thresheree, one competitor stood out: Cyrus Johnston. A large burly man with a big beard, he wore a Stihl ball cap, a yellow T-shirt, cargo shorts, and work boots. Over his bare legs he'd clipped on saw chaps, and he had mounted safety side shields onto his prescription glasses. Johnston's wife, Leisa, competed in the smaller engine classes, and he ran the bigger saws. He stood before the log in a more athletic stance than the other competitors. This was noteworthy. Some people handle chainsaws like they are holding onto the tail of a snake that could flip around and bite them. But extending the saw away from you makes it more difficult to control. Johnston's feet were in an orthodox boxer's stance, and he held the saw close to his body near his center of gravity. He had total control of his saw.

Even though his saw's chain slipped off at the very end of his last cut, he still tied for first place at 5.3 seconds. The judges conferred. "The consensus was to let Cyrus saw again," the announcer explained over the background rumble of warming saws. "That's something with these local contests. You don't have a comprehensive list of rules. It's just for fun, yet it is competitive. We discussed it and the people involved agreed to have Cyrus run again. He has to put a different chain on because, as many of you probably know, when you have a chain derail like that, the drivers get beat up on it

The bent-leg, locked-solid athletic stance of Cyrus Johnston gave him a solid center of gravity to make his event's three required cuts at the Union Thresheree in Symco, Wisconsin.

and they don't fit down in the rail of the bar. You can derail a chain on that upcut. I have seen it happen. It's not often, but if you're still moving over as you're going up into the wood, that can tip the chain."

Cyrus derailed on another cut. "The sprocket must be bad, but I made it through," he said aloud after cutting his engine. Despite the setbacks, he ended up taking first, second, and third place with three different saws in the 100-cc-and-under class. His first-place cuts were made in 3.8 seconds with a Dolmar 7900, a professional-grade saw known for producing great power for its weight. In 2003, the Dolmar made three hundred limited edition models of the 7900 with black trim and boosted power output. The first model was not for sale, as it was awarded to the winner of the

German Logging Championship, a competition described as the Formula 1 of forestry skills.

Johnston lives in Menominee in Michigan's Upper Peninsula, where he owns Johnston Performance Chainsaw & Small Engine Repair. I contacted him a couple of months after the Union Thresheree and asked if I could visit his workshop—a garage-sized building behind his house on the edge of town. On one side of the garage is a garden full of huge peppers and on the other side of the garage are logs for testing his chainsaws.

"I up-classed all of my saws and I won almost everything that day," Johnston explained to me. "Every one of those saws is modified." By up-classed, he meant he was purposefully running a smaller-sized saw against bigger saws—like a cruiserweight moving up to heavyweight to gain a shot at another boxing title.

Johnston specializes in porting chainsaws, modifying a saw's exhaust, intake, or transfer ports to increase its horsepower. If you were to take your stock 50 cc chainsaw to Johnston to have it ported, you would get back a 50 cc chainsaw that cuts like a 70 cc chainsaw. You would never know it was modified, even if you popped off the top cover to look at the engine. The adjustments he makes are so small that just a handful of people in the chainsaw world can notice the difference. In some cutting competitions, a chainsaw must look like a stock chainsaw. For the hot saw competitions, it can look like anything. Johnston's ported saws are like the cars in the *Fast & Furious* films—on the street they look like normal cars, but under the hood, they are modified for incredible performance.

Johnston grew up in Middle Inlet, Wisconsin, and fell in love with the sound of his father's Jonsered chainsaw. "The sound of that engine was different than anything else I had ever heard and that just stuck with me over the years," he recalled. His father was an industrial mechanic at Ansul, a manufacturer of fire extinguishers and fire suppression systems. It was the 1980s and his dad often took him to work. Johnston admits things were different back then, with fewer OSHA regulations, and his father taught him how to weld in the maintenance shop when he was eight years old. His dad said young Cyrus welded better than most men who worked there. He had a gift for understanding mechanical operations, and as a teenager, he worked on three-wheelers, dirt bikes, snowmobiles, cars, and trucks.

Johnston became a machinist and a certified pipe welder at Marinette

Marine, a shipbuilder for the US Navy and Coast Guard. For years, he worked on chainsaws on the side, and his customers liked what he did with them. "I've always been the guy that takes that motor and makes more horsepower," he said. "Way back when, you didn't go to school to learn how to modify a car engine. You just did it. Now you can go to school for that. There is no school for modifying two-stroke chainsaw engines. The guys who know a lot about it, who wrote books on it, are the Japanese bike makers. They are probably the oldest ones doing it. A lot that I learned was from books written about Kawasakis and Hondas and guys playing with them."

Eventually, Johnston quit his job in order to focus on chainsaw work. He had concerns about the loss of a steady paycheck and benefits, but his name was out there in the chainsaw world, and he kept busy in his workshop. His lucky break came when an older truck driver drove by his property, saw his front-yard sign, and stopped in to inquire about a modified saw. The man wanted to use the saw for the opening ceremony of the Great Lakes Logging & Heavy Equipment Expo. At this three-day event, also called the Logging Congress, instead of cutting a ribbon for the opening ceremony, the organizers cut a log. The man invited Johnston to help him set up for the 2022 expo in Green Bay, where two expo representatives ran his 1948 Mercury two-man saw for the opening ceremony. Johnston attended the next year's expo in Escanaba, Michigan, and this time, Johnston brought his ported saws. He got them into the hands of some industry players, then he made some videos and posted them online to get the word out about his saws.

Now, he ships his saws as far as Washington state and Canada. Many go to Minnesota. "I've had random people walk into my shop from Texas and later send me saws. 'I saw your sign and I stopped. You build hot saws?' Well, yes, I do," laughed Johnston. "It's just meant to be." He charges $1,000 per saw for his porting and modification work. "There are a lot of places where you can get it done a lot cheaper than that," he admitted, "but I am fascinated by saws. It's the most powerful handheld device there is, and so I make them hot. The loggers have no problem paying that price because they want something hotter than anything they have ever run and that's what I build."

At any given time, Johnston has four types of saws in his workshop.

First, there are the saws with problems brought in by local cutters who need to get back to cutting firewood. Then there are the saws that Johnston has collected over the years—both antique and modern ones, hanging from the ceiling and stacked on shelves. Among these, there are two types: those that may be restored and those that might be used for parts. Finally, there are the saws sent to Johnston by professionals for his porting work—these are the hot saws, the custom-crafted saws rebuilt for speed. These are the standouts.

Whereas a car engine has a cam shaft that controls the timing of the internal processes, a chainsaw engine is a piston-ported two-stroke. The position of the piston in the cylinder is what controls the timing of the processes. And the timing of the processes is the time it takes for the intake charge to come into the engine, the time it takes for the charge to go into the combustion chamber, and the time during which the engine is under compression. This is all controlled by the position of the piston in the cylinder.

"A chainsaw two-stroke is just a crappy air pump," Johnston explained. "All it does is pump air. What I am doing is making it a more efficient air pump by enlarging the ports and changing the timing of the ports—how long this port is open versus how another port is opening or closing. More air improves efficiency. The more oxygen you can get into the cylinder, the more fuel you can get into the cylinder, and that's what makes horsepower."

What is most visibly noticeable about a hot saw is its expansion chamber pipe—an exhaust pipe that widens into a chamber with cones on each end. As the exhaust is expelled, it hits the divergence cone, goes into the large chamber known as the belly, and then hits a diffuser cone and comes through a tiny pipe at the top called a stinger. Johnston's expansion chamber pipes are made out of sheet metal, but the intense heat of the exhaust gives them a burned coppery patina.

One of the design flaws of chainsaws, Johnston explained to me, is that "they must be made extremely compact, but they are extremely inefficient engines. Power-to-weight, they are a great engine, but they are very inefficient because they waste fuel." In a typical two-stroke engine, about half of the air and gasoline mixture that enters the engine during intake comes out in the exhaust.

The expansion chamber pipe is designed to rectify this wastefulness through a process like recycling. "A two-stroke sends out what is called the pulse wave," Johnston said. "The pulse wave comes out in this [expansion chamber] pipe and the pulse wave is amplified. When it hits this [diffuser] cone, it sends this wave back. It forces that fuel back into the engine. . . . This does two things. This pipe will actually suck more fuel through the engine and into this pipe, and it will ram it back into the motor."

Just outside of his workshop, Johnston showed me his testing area. When one of his pipes doesn't work well on a particular saw, he has to cut it apart and rebuild it. "This one has what I call a slider dog," he said, referring to a mounted metal bar. "It's to protect the pipe. When you're racing, it's an adrenaline rush. You don't want to smash into this pipe and bust the cylinder off." He picked up a pipe. "This cylinder is dead because this broke off," he explained. "Ka-bang and she was done."

Building pipes is time-consuming and rebuilding them because they cracked or broke or snapped is even more of a pain. Sometimes pipes need

Parker Jameson ripped through the wood with his hot saw at the 2023 Union Thresheree. The exhaust pipe—made of two cones joined together to create an expansion chamber—was designed to maximize power for two-stroke engines.

to be rebuilt simply because all of the variables did not come together in what Johnston refers to as a symphony. There is a science behind these modifications gleaned from Nazi rocket testing and Japanese dirt bike aficionados, but also it is an art and a craft to get it all right. The pipe can be too short, the chamber can be too fat, and the angles of the cones need to be just right in order to amplify the pulse wave and send it back at the exact right time. The only way to test these factors is to run the saws and cut into some wood. When all of these things come together, Johnston calls the experience getting "on the pipe."

"You just hear it," he told me. "It makes a different sound and you know you're on the pipe. When it falls off the pipe, it doesn't make any power." He picked up a hot saw. "When I run this saw with this pipe, I can feel it on the pipe, but as soon as I hit the wood—if I touch too hard—it falls right off the pipe and the motor falls on its face. The wood doesn't lie."

Another difference between stock saws and hot saws is the type of fuel they run on. Some of the sawyers at the Union Thresheree were running nitromethane, or "nitro," in their saws. In the automotive world, hot rodders have experimented with this explosive compound for decades as a fuel additive that provides greater power. It has also been used for rocket fuel, and its volatility can blow out engines. Nitromethane doubles a machine's horsepower. According to Johnston, if a person tried to use nitro in a stock saw, it wouldn't run very well or possibly at all. Doubling an engine's horsepower requires doubling the amount of fuel, and most stock chainsaw carburetors cannot deliver that much power.

Johnston pulled out an 80 cc nitromethane saw for me to observe. Someone else built it in the 1980s, and Johnston acquired it and restored it. The original owner had used a carburetor from a snowmobile and had removed the chain brake. Typically race saws don't have chain brakes, but sometimes they are required in competitions for stock-appearing saws. "You don't want to engage that," Johnston explained. "Then you lose. At one race we were at with my stock saws, we started with the saw on the ground and hands on the log. I did it five times. I reached down and grabbed the chain brake instead of the handle."

Chainsaws can run on aviation gas and racing fuel, but the problem is that these are usually stored in a vented tank that allows the atmosphere to get inside the tank, which dilutes the fuel. Johnston walked over to a shelf holding cans of 108-octane and 117-octane racing fuel. "The first time you open that can," he said, "it's the purest fuel you're going to have. As soon as air gets in this can you're losing power because the air has moisture in it, and that moisture goes straight into the fuel and then the hydrocarbons of the fuel are evaporating."

The higher a fuel's octane, the more compression is needed from the engine. If the saw doesn't have a high-compression engine, the fuel will burn too slowly and the saw will be less powerful. High-octane fuel is meant for high-compression engines, because the more the fuel is compressed, the more volatile it gets, and it will explode by itself. "If you have high compression and you ran [a relatively low] 87-octane in it," Johnston explained, "the piston is traveling up and it could ignite the charge just by compression, which could blow the piston down on the wrong stroke and totally annihilate everything. It's called detonation. Some people call it preignition."

Ethanol is terrible for chainsaws and provides Johnston with much repair business. "The fuel companies will tell you that gasoline with ethanol in it is good for twenty-eight days," he said. "After that the ethanol starts separating out of the fuel. I can tell you from experience being a mechanic, it does not last that long. When it separates, ethanol attracts moisture. Most oils do not mix well with ethanol. You will have carburetor issues. If you leave that gas sitting in there over the winter, your saw is not going to start in the spring. I deal with about sixty of them a year that are like that."

The air filter is another part that Johnston frequently replaces or modifies. He showed me a saw with a two-stage foam filter commonly used on dirt bikes. Its surface area was much larger than that of a factory air filter, so it allowed more air to come in. On his nitro saws, however, Johnston doesn't use filters at all, since the saws need so much air. Without filters, they can suck in some wood debris and they can also break down more quickly, something Johnston calls "going to grenade."

"This is a 1983 Jonsered," he said. "And parts are very hard to find. I spent a lot of money getting the parts I needed, so I'm running an air filter so it will lengthen the time that I have with this saw. This is a pipe that is actually off of a dirt bike that I modified to work on this saw. I grenaded this saw once and then I had to hunt down another cylinder that I found in Canada. It's easier to work on a newer saw so you can go and get parts for it. When you're playing with old stuff, these older saws are more fun. Half the fun is finding all the stuff and then building the thing."

Johnston is so detailed in his saw alterations that he even makes changes to the gears. He demonstrated this by walking to a drawer, plucking out some loose gears, and returning with a jingling handful. "That's a seven. Here's an eight. Here's a nine," he said, showing me the most common sizes for off-the-shelf saws. He tossed the coin-sized drive sprockets onto the workbench. "This is a ten. Now, when you get into the nine and the ten, you need power. This one here has a fifteen on it. Most stock saws will not run a ten unless you're going to cut twigs with it," he said.

The job of a sprocket is to take the energy created by the two-stroke engine and transfer it to the bar to make the chain go around. There are two types of chainsaw gears: spur and rim sprockets. The small round rings Johnston produced were rim sprockets. This type is the one most commonly used, since they are easier and less expensive to replace. They are composed of two discs with channels set around the circle. The star-shaped spur sprockets are considered more durable as they are one solid part. A sprocket will wear out over time, which can lead to the chain derailing off the bar.

Like gears on a bicycle, the sprockets on a chainsaw can be changed for optimal performance. Gearing up or down is like shifting gears on a bike: the gear for pedaling uphill is different from the one used on a straightaway. For cutting softwoods, speed can be increased by increasing the sprocket's diameter. For cutting hardwoods, more torque is needed, and a smaller diameter is desirable. The gauge and pitch of the bar and chain must match the sprocket. For most sawyers, the manufacturer's sprocket is all that it needed. Having an identical replacement in a traveling pouch of chainsaw tools is always a good idea.

For his hot saws, Johnston uses outsized sprockets. Replacing standard sprockets with larger ones, though, often means that he must modify other parts of the chainsaw. He showed me a saw with a thirty-six-inch bar that he had modified at the end. In order to get the chain to ride from the bar to the sprocket, he had to precision-cut it.

"You can't buy a bar to run a ten-pin sprocket," he explained, "so the bar has to be modified to run the bigger sprockets, and the bigger sprocket increases chain speed. By porting a saw, you can change the rpms at which the engine operates. A regular saw got probably about 9,000 rpms at peak power. Now this saw when I ported it, it now is probably 12,000 rpms at peak power. But this is a race saw and that won't be any good for working in big timber. It's a symphony. You increase rpms, but you lose a little end torque. So, when you're trying to cut really big stuff, you lose your bottom end. [It's like how] a semitruck has a lot of bottom end to pull all that weight, but a drag car has a lot of top end to go really fast. It's the same thing with chainsaws."

The final component that Johnston often modifies, and one of the most important, is the chain. Some competitions have rules on chains. Some require round-ground chains only. Others allow square-ground chains. These terms refer to the shape of the cutting teeth. A round ground is easy to sharpen, stays sharp longer, and is more durable. It can also be easily hand sharpened with a round file. A square ground has square teeth; it cuts fast and is the most efficient of all tooth shapes, but it dulls faster and is trickier to sharpen, requiring a special file or grinder. A mix of the two is the square-tooth round ground. This version is faster than a round-ground chain. And, although it dulls quickly, it can be sharpened with a round file.

For general use, Johnston uses square-ground chains, as they cut faster. He will use a round-ground chain only if he is in a hurry and one happens to be lying around. His race chains are modified for high speed and low drag. Their teeth are sharp enough to draw blood from a careless touch. Johnston files down the back of his saw teeth at a chamfered or beveled angle to allow for more chip clearance. He files the front of the rakers and sets them at a depth comparable to those on a working chain. And he files the drive links—the teeth on the bottom of the chain that fit into the groove of the bar and encircle the sprocket—until they are smooth. All of these efforts lighten the chain but greatly reduce its strength. These are not working chains, as they are used only a few times a year on race days. The chains are honed and streamlined to make the fastest cuts in the world.

"Modifying saws requires a lot of testing. Test and test and test and document and test," said Johnston. He once took two identical chains out of the box and modified one into a race chain. He filmed himself using both chains on a ten-inch log. Using editing software, he could watch the cutting in slow motion at thirty frames per second. He timed both cuts. The race chain cut 1.6 seconds faster than the stock chain.

A sawyer usually tests the tension of a chain by pinching it and lifting it up. Any looseness is easily felt and can be adjusted. Most sawyers like to feel a little snap in the chain. In contrast, the chains of hot saws are much tighter, especially on saws with large drive sprockets. Loose chains can derail and go flying. And on modified and custom saws, the chain is thrown off the bar with much more force than on stock models.

"I had that derailment [at the Union Thresheree]," Johnston admitted, "and I don't know if anyone saw it, but that chain came around the back around my leg and almost hit me in the butt. I got a cut in my chaps from that, and that's what slowed it down enough to not cut the back of my leg. But that saw is over ten horsepower. The chain was too loose. That was the problem. All of this stuff is like a symphony. If you have one piece that is out of tune, that symphony sounds like crap and so does a two-stroke engine. Everything works together. You can't go to school to learn this stuff and the consequence of that is you break shit."

7

Cutting at the Fire's Edge

The Saw as a Maker of Habitat

I run the chainsaw until it burps, a sign of low gas, then I make one last cut and kill the engine of Katana, my nickname for my Husqvarna 550 XP. Since I am averse to my saw's chain touching the ground, I place the tip of the twenty-inch bar on my steel-toe boot and rest my arm on the saw as if it were a walking stick. I survey my morning's work of butchery and destruction. I call this the Battle of the Four Armies: me (an army of one) versus buckthorn, honeysuckle, and autumn olive. Katana is my main weapon in this endless campaign.

My enemies are nonnative invasive species that have overtaken the forests of Wisconsin. These three species are half shrub and half tree, and they create an understory canopy that hogs most of the sunlight before it reaches the forest floor, leaving the native plants struggling for survival. The nonnative plants host far fewer insects than native plants, and this in turn means less forage for turkey poults, ruffed grouse, and songbirds. Some of the patches are so thick that I must crawl on my hands and knees, scoot the running chainsaw forward, make a cut at the stump, push the brush aside, and crawl to the next one.

There is also a smaller battle that I wage against Japanese barberry. It is a prickly shrub but it doesn't get much over waist high, so I am less concerned about it than I am about the other ones that grow much larger. It does, however, colonize the woodscape quickly and provide a favorable habitat for wood ticks. Some believe that the spread of Japanese barberry is linked to the spread of Lyme disease. I've been known to burn a whole tank of gas just to destroy this foul plant. Its cut stems reveal a bright golden color, and I like looking over a hillside of slashed barberry and seeing their putrid yellow remains. It is the sight of victory.

During the COVID-19 pandemic, I did some line-clearance tree trimming on a right-of-way in Fitchburg, Wisconsin. I spent hours bear-crawling under the honeysuckle and cutting the multistemmed trunks low to the ground. Once enough were cut, I killed the saw and pitched the shrublike trees down a steep slope to later be fed into a wood chipper. This right-of-way crossed a jogging and cycling path that was overgrown with invasive honeysuckle. It's what urban planners like to call a "greenway." Some may believe the honeysuckle is an eco-friendly feature of the community, but I see it as an ecological hellscape of monolithic invasive brush.

I hiked over to my backpack, fetched two cans of Leinenkugel's, and stabbed them into a melting pile of snow. They would be my reward for this workout. I topped off the saw with fuel and oil, and I gathered the freshly cut honeysuckle, buckthorn, and autumn olive into huge piles the size of ice-fishing shacks. I mulched them down with my chainsaw, making a salad of woody debris. I cut down through the conjoined branch collars and severed as many limbs as possible in one swipe of the bar. I diced the lateral branches into twigs. Eventually, the once-enormous brush piles were leveled to knee height. Saw mulching is aggressive, fast, and physical work. When I do it, my saw and my body work in an automatic and instinctive trance.

The operator of a chainsaw can be the shepherd of a forest, a tender to the trees. Healthy woodlands are necessary for the contentment of the human soul. Like many other living systems, forests are constantly being invaded by outside organisms. I take great pleasure in destroying these life-forms. After I reduce certain invasive species to a pile of stubby sticks, they eventually decompose and enrich the soil but not before providing "rabbitat," a home to the small animals of the forest. Once I finally finished my battle that day, my underlayer was soaked with sweat and I cracked open one of my Leinie's.

Back when I acquired the property where I currently live, I was aware of a section of five acres that were overgrown with invasives. After a few years, I launched my first campaign: total destruction of the plants' method of sexual reproduction. Buckthorn, honeysuckle, and autumn

Buckthorn stays green after other trees have changed colors, which makes it easy to identify for eradication as an invasive species. The berry clusters are another identifying characteristic.

olive all reproduce by berries. If I cut down the large mature plants, I figured, there would be no more berries. However, berries from years past can lay dormant in the soil and spring to life next year. Berries can also be transferred from a neighbor's unchecked property by birds crapping them onto my land.

Additionally, the cut stumps will shoot up suckers, and they always come up thornier than before—a vengeful defense mechanism. The suckers on the larger stumps can grow to chest height in a year. I always cut them down a year later. One stump took six years of annual cuttings until it finally used up all of its stored energy, withered, and died. On some stumps, I use Garlon 4 Ultra, an expensive and potent herbicide. I use it sparingly: just a quick squirt from a handheld bottle. The liquid is a metallic blue-purple color and has a foul, unnatural smell. Once it's applied to a freshly cut stump, death is certain.

I recalled a quote from Aldo Leopold: "One of the penalties of an ecological education is that one lives alone in a world of wounds. Much of the damage inflicted on land is quite invisible to laymen." Many people cannot identify buckthorn, honeysuckle, and autumn olive, but once you learn to recognize them, you see them everywhere. To friends and neighbors, I preach the gospel of identification and annihilation and cajole them to pick up their axes, loppers, and chainsaws to rid the land of these forest-choking invaders. I give the invasives some competition by planting hundreds of hazelnut, pine, and spruce seedlings every spring.

The hill where I clear these invasive species is part of a swarm of drumlins—a group of teardrop-shaped hills sculpted by ice sheets during the Wisconsin glaciation. My property is on the terminal moraine where the last glacier to cover North America stopped advancing. The Ice Age National Scenic Trail that marks this ancient geological delineation is a few miles from my house. The trail was recently designated as a unit of the National Park Service and is maintained by a small army of volunteers. Some of them are trained to operate chainsaws for cutting windthrow trees that block the trail.

At this location, a subtle change occurs in the Wisconsin landscape. To the north is the Forest Transition eco-region—the gateway to the great North Woods. And to the south and west are two related eco-regions, the Central Sand Plains and the Central Sand Hills. These two eco-regions meet at the northeastern tip of Sauk County, where Leopold wrote his masterpiece, *A Sand County Almanac*. The first two sentences convey the heart of the book: "There are some who can live without wild things, and some who cannot. These essays are the delights and dilemmas of one who cannot."

For me, the *Almanac* is a "dipping book," one of those books you read cover-to-cover once or twice and then, for the rest of your life, dip into now and then to refresh your mind with its ideas. It is a book of essays, a philosophical tome, a nature journal, a record of seasonal change, a compilation of wildlife and botanical observations, an eco-ethics primer, a study of wilderness, a diary of cabin living, and a template for environmentalism before the word was invented. Although many of the essays take place in Wisconsin, it could also be described as a collection of travel writing, as Leopold journeys into Illinois, Iowa, Arizona, New Mexico, Oregon, Utah, Chihuahua, Sonora, and Manitoba. It is also an instructional handbook on wise and prudent living. To wit: "You do not annex a hobby, the hobby annexes you. To prescribe a hobby would be dangerously akin to prescribing a wife—with about the same probability of a happy outcome."

Wisconsin does not have a county by the name of Sand, as the book's title implies. This fictional name is inspired by the eco-regions of the Central Sand Plains and the Central Sand Hills, an area that was once partly covered by a 1.1 million–acre glacial lake. The soil of this area is sandy and poor for most kinds of agriculture—except for the cranberry industry, which straight-lined the bogs for production. The region contains marshes, meadows, pine barrens, oak savannas, small lakes, cold-water creeks, remnant and restored patches of tallgrass prairie, and sandstone cliffs and gorges unlike any other in the state.

In his essay "Good Oak," Aldo Leopold describes sawing an old oak tree with a crosscut saw. Throughout the essay, he includes the refrain: "Rest! cries the chief sawyer, and we pause for breath." Leopold would not have used this narrative technique had he done his work with a chainsaw, as he would have made the cut in less than a minute. UNIVERSITY OF WISCONSIN–MADISON ARCHIVES, COURTESY OF THE ALDO LEOPOLD FOUNDATION

It is in this area that Leopold, who was then a professor at the University of Wisconsin–Madison, bought eighty acres in 1935. On the land was a chicken coop, which he turned into a cabin. Known then—and now—as the Shack, it was a place he considered "a refuge from too much modernity."

In 2017, I visited the Shack, which is now preserved by the Leopold Foundation, a nonprofit group established by Leopold's wife, Estella, and their five children in 1982. The foundation's mission is to advance Leopold's "land ethic," a philosophical concept from an essay of the same name in the *Almanac*. The land ethic encourages people to interact with the land as good stewards, and Leopold uses an all-encompassing definition of *land* that includes birds, bees, trees, deer, dirt, water, fish, and everything else in nature.

The foundation's goal is to restore wetlands, prairie, and savannas for the survival of grassland bird species on the land surrounding its headquarters and the Shack. The decline of grasslands in the southern half of Wisconsin is a tragedy. Few people think of Wisconsin as a prairie state, but much of the landscape was grassland before the arrival of Europeans, when the area was home to the Sauk, Meskwaki, Ho-Chunk, Kickapoo, and Menominee people. French voyageurs later used *prairie* (the French word for *meadow*) to describe the land.

The foundation uses the diversity and abundance of birds on its land to judge the overall health of the landscape. When the ecosystem has good soil and water and a diverse mix of plants, the birds appear. It is similar to the canary in a coal mine metaphor, based on miners who would bring a caged bird into a mine shaft to determine if dangerous gases were present. If the canary keeled over, it would give warning to the miners to make their escape. Similarly, an unhealthy prairie will attract and sustain just a few species of birds and probably not many birds in general. It should be a warning to us all.

In its effort to transform this landscape back to prairies and savannas, as it was in the days before European colonization, the foundation is challenged by the fact that much of its land is covered with forest. Even on the land that *has* been restored to prairie or savanna, tree seedlings (both native and nonnative) creep in and take root in a landscape where they were never supposed to appear. The trees are part of a relentless slow-motion drive to revert the land back to forest. This means the chainsaw is a critical tool for the foundation's mission.

On a snowless mid-December day in 2023, I met Arik Duhr, the site manager for the foundation's property and the nearby private properties partnering with the foundation for its conservation mission. He greeted me wearing Pfanner saw pants, a hooded sweatshirt, and a stocking cap. Those three items are like an unofficial uniform in the world of tree work. We hopped into his Ford work truck for a tour. As we pulled out on Levee Road, I commented on a nearby house with a portable lumber mill outside. It turned out to be Duhr's place, and his mill was a TimberKing. He said he also milled wood with a Granberg chainsaw mill.

As we talked shop about mills, chainsaws, and woodworking, he slowed the truck down and slowly backed up. "This is what everything we just drove by looked like," he said, gesturing at the landscape beyond the windshield. He was referring to what the landscape looked like before the foundation's restoration work. "This is an in-holding. We don't own it. But it shows you what no-management looks like. It's a drastic change."

The contrast was stark. In front of us were brush-choked woods, but to the sides of us was an open-air restored savanna dotted with oaks. The delineation was a property line, but it was as if an invisible wall separated the two types of ecosystems. The ground on the savanna was scorched and blackened, but it had once looked like the thick woods before us.

"We had a November fall burn, roughly 180 acres," Duhr explained. "Prescribed fire on a larger scale is one of the few things that have a chance for us getting ahead. We're on 4,500 acres. The amount of invasives and woody propagules—like a shrub or a bush or a tree—for a grassland or a savanna, those are the enemies." He edged the truck up to a stand of young aspen that had resprouted and would not die despite undergoing multiple burns. The young trees were as tall as me, with trunks as thick as mop handles. Prescribed burning only temporarily pauses regrowth. Duhr said getting rid of the aspens would require "mechanical intervention," meaning a crew member would eventually lay waste to the undying aspen with a chainsaw or run them over with a skid steer and a Fecon mower head—essentially a giant lawnmower for woody vegetation.

Duhr explained that the area used to be a floodplain forest, and the sandy banks of the Wisconsin River were just a short hike away. The Kilbourn Dam in Wisconsin Dells, finished in 1909, was the first large hydroelectric dam on the Wisconsin River. Upstream, it flooded some of the area's sandstone caves, canyons, and cliffs—the natural wonders that make Wisconsin Dells famous. Downstream, it stopped high-water events that would have scoured the soil of the floodplain, keeping forest growth in check.

We rolled up to a four-person work crew set up on the edge where savanna met forest. Duhr dropped off a box of tree-marking spray paint. A couple Stihl saws lay on the ground. Blue paint is for the "keep trees" of white oak, burr oak, and hickory. Orange paint is for the "take trees." There is no standardized forestry color scheme for marking trees, but there is a standard for marking them at chest height so they are easy to see.

Private operations and state and national foresters all use their own color combinations.

The foundation crew runs mostly Stihl saws, with one Husqvarna. Duhr's favorite saw is a Stihl 261 with a sixteen-inch bar. "I love the power, the weight, and I can use it all day and not feel it," he said. "I have bigger saws. I have an 880, so I have the other end of the spectrum." The Stihl MS 880 is the most powerful series-produced chainsaw in the world. Its engine displacement is 121.6 cc, and the saw will accept guide bars that are almost five feet long. It's a saw built for cutting big wood.

Part of Duhr's crew is made up of fellows: recent college graduates who have an interest in the foundation's work. No matter their field of study, they all go through a chainsaw safety program and get hands-on work experience in the field. That day, he had them doing some light saw work and marking trees as keepers or takers. Duhr explained the nuance of their work: "You see these woods and a lot of people would be like, 'That's fine. It's just woods, right?' But we know it doesn't do what we want for birds. The forbs and grasses that grow underneath it don't thrive because it's so shaded. This ecological community will be an oak woodland. We prioritize grassland and savanna. Woodland would be about 60 percent canopy cover. Savanna is more like 30 percent canopy cover. This spot, we would never push to be savanna because the structure is not there, but it is there for an oak woodland. I can see good-quality oaks capturing 60 percent of the canopy."

Reducing a section of overgrown woods to 30 percent tree canopy to create an oak savanna means an enormous number of tree removals on 4,500 acres. It's too much for Duhr's crew, so eventually, they'll bring in a commercial logger. But not before Duhr and his crew take a first crack at it. With chainsaws, they cut and remove small trees around ten inches DBH (diameter at breast height), and later Duhr will mow the existing bramble and brush with the Fecon-mounted skid steer.

After that, he explained, "we can have a logger come in and see it clearly: 'I see the volume of keep trees, I see the volume of take trees, and I'm going to give you this bid on all of the orange and do all the work.' And he or she is going to be stoked because they don't have to drive through everything. Our work opens it up a ton." If they do not get a bid from a logger, then they must do it all themselves.

Duhr is a professional arborist certified by the International Society of Arboriculture, but before receiving that certification, he studied kinesiology and worked for eleven years in health clubs and clinics. He made a career change after deciding that trees were easier to work with than people. For a time, he worked for the city of Madison. He once felled a silver maple that was hollow, and when the main stem was cut, by his estimate five hundred gallons of water poured out. He later worked at a tree care company and managed a tree farm before joining up with the foundation.

"So much of the work we do is removing vertical barriers and creating more habitat continuity," he said. "You see grassland here, but then you see a forest. For our priority bird species, they will fly all day long out here [over the grassland], but they see that barrier [of woods] and they don't think there is anything of value behind it. We're making management goals that people think of from a land preservation standpoint: 'You're cutting trees? That doesn't make sense!' But we have such a unique opportunity because grasslands birds are really in peril."

We drove down the two-track where the edge of the savanna and the edge of the forest met on a bowl-shaped hillside that looked like a scorched-earth war zone. It used to be 4.5 acres of nothing but black locust. In 2002, black locust had been harvested as timber in the same spot. With the canopy gone, the dormant locust seeds exploded with new growth and the stumps resprouted. The foundation signed on to a government program allowing the US Fish and Wildlife Service to reseed it as grassland; however, that required removing the new pole-sized locusts. The stump-sprouted trees were five to eight inches thick, and the seedlings were three inches thick. It was all chainsaw work. Duhr estimates they invested four hundred hours into the spot. They piled up the pole-sized timber to be hauled off with a grapple-equipped skid steer.

"What you don't see there is about $5,000 of prairie seeding that's gone in," Duhr told me. "We planted it a month and a half ago. Come back in a year and it's going to look like a lot of annual weeds. Come back in two years and it will still look like some annual weeds but you'll start to see some good stuff. Come back in four or five years and we hope to have bee balm and black-eyed Susans and cup plants. We'll manage it with fire. It's part of a burn unit that we will work to keep the invasives in check."

Duhr pointed out a wooded ridge in the distance. In the middle of

The shadow of a lone oak tree at the Aldo Leopold Foundation stretches across a barren landscape that once held a thicket of black locust—all removed by hand cutting with chainsaws. Although it looks grim now, it will turn into a savanna of wildflowers and native prairie plants.

the ridge was a bare-looking spot dotted with a few small oak trees. He called the spot a "punch-through." Birds and pollinators were living in the savanna on the near side of the ridge but not going over the ridge to the prairie on the other side because the wooded area between them was too much of a barrier. During bird surveys, foundation workers would find Henslow's sparrows on one side of the ridge but not on the other. "Nothing could get there," Duhr explained. "We would do bird and pollinator surveys and nothing was in [that prairie on the other side]. This last year we put in the punch-through over the ridge. We're really excited. I'm not a betting man, but I'm guessing you're going to see stuff in there we have never seen before because now it is connected."

Duhr continued, "The use of chainsaws is the only way to get us back so we can maintain [the land] with fire. After the glaciers receded, you had a fairly consistent fire interval. You would have fires on parts of the landscape almost every three or four years here." He pointed out some small scrubby oaks. “This is red oak regrowth. If it’s burned every three or four years, it just goes away. It cannot keep resprouting, especially with large herbivores browsing it. [Hundreds of years ago], we had elk and bison that would keep it in check, and fire would set it back.” Now, even though deer live in the area in greater numbers than before, they don’t eat the red oak regrowth because they have such easy access to nearby corn and soybean fields.

The foundation’s mission is to create habitat for high-priority grassland bird species, but landowners can use the chainsaw in a variety of ways to benefit other birds and wildlife. Ruffed grouse, for example, thrive in a young aspen forest with high stem density—exactly like what remains after a clear-cut. Awash in sunlight with the older trees removed, suckers and new sprouts spring alive by the thousands. Grouse need this sapling transition as it provides vertical cover from raptors and ground predators. Walking through one of these stands is like squeezing through the bars of a jail cell. A patchwork of small clear-cuts through an aspen forest made over the course of years will provide ideal habitat for the grouse. As one cutover matures, the grouse move to the younger clear-cut.

If the ruffed grouse is the king of gamebirds in northern forests, the bobwhite quail is the southern prince. It is a grassland bird—also, a delicious one. In the autumn 1992 edition of *Esquire Sportsman*, Jimmy Buffett wrote an essay about quail hunting titled, “Everything in the Woods Wants to Eat a Quail (Including Me).” The title is reflective of how the bobwhite quail is so heavily preyed upon. When a meadow of native grasses, the ideal home for quail, is surrounded by a few large dead trees, raptors perch upon those bare branches and deadeye the meadow looking for an easy lunch. To help out the quail, you just need to fire up your chainsaw and fell the dead trees. Then, the raptors cannot ambush from above. Such a simple solution has a big impact on bobwhite quail survival.

Hunters can create habitat for whitetail deer by hinge cutting, a technique where the sawyer cuts halfway through the trunk (no notches, just a straight cut), causing the tree to hinge over to the ground but still stay alive. The cambium layer on the uncut portion of the tree will allow the tree to keep growing. This creates a ground-level food source for the deer and also provides bedding and escape cover. This habitat creation technique has some downsides, however, and one is that invasive species can move into the new patch of sunlight.

At one time, buckthorn was the foundation's biggest worry when it came to invasive species. But after years of chainsaw cutting, mowing, and prescribed burns, the buckthorn is mostly under control. Now the biggest concern is spotted knapweed, a noxious weed that can withstand fires. The foundation also has barberry, honeysuckle, and black locust to contend with.

Back on Levee Road, Duhr and I drove out through a restored savanna dotted with oaks—an area that was once commercially harvested by loggers. "Now, we've done some Fecon work here. There is too much acreage here for chainsaw use, so it's mostly edge cleanups." But not all of it was cleaned up. Next to the two-track was a patch of buckthorn with stems about as thick as my wrist. "This is a stand of buckthorn I'm on the hook for for the next calendar year," Duhr said. "It's about eleven acres. It needs to be removed for a federal program called EQIP." The Environmental Quality Incentives Program, or EQIP, is a conservation program through the Natural Resources Conservation Service, a branch of the US Department of Agriculture, that works with private landowners on conservation projects.

Duhr described the work that needed to be done in the area. Mixed in with the impenetrable stand of buckthorn was some swamp white oak, a key "keep tree." Before blasting in with the Fecon mower, sawyers would come in with a chainsaw and hand cut around the white oak. If the Fecon mower head were to brush against the tree, it could be gouged or even killed. I was familiar with this protocol from line-clearance arbor work on right-of-ways. We were instructed to hand cut a buffer area around all structures. If a Fecon mower were to strike a wooden utility pole supporting a high-voltage line, it could be disastrous. If the structure was metal, the mower head could be severely damaged. Similarly, when we worked on slopes where the skid-steer mower could not reach, all of the woody vegetation needed to be cut out by hand with chainsaws.

We left the truck and hiked out to a ground blind that overlooked the Wisconsin River. The foundation hosts sandhill crane viewing programs from November through early December, when thousands of cranes gather on the sandbars before flying south for the winter. There were no cranes to be seen at that time, as they were feeding out in the farm fields. On the drive over, I had seen hundreds of them within a few miles of the foundation's property.

Duhr pointed out an island in the river and told me it was one of his favorite views. The island was once completely covered with trees. To transform the habitat, he and his crews loaded their chainsaws and gear into canoes and rowboats, crossed the channel, and spent days working on the island. They left three trees standing, evenly spaced apart: a basswood, a river birch, and a cottonwood. With the forest cover removed, the sandhill cranes now flocked to the island to roost. When the river freezes over, the island is no longer a sanctuary, as coyotes can cross the ice, and by that time the cranes have started their southward migration.

As we drove back to the headquarters, Duhr confessed that there had been a chainsaw mishap during a prescribed burn two weeks earlier. When doing prescribed burns, his crew always has chainsaws at the ready for the moment the wind changes.

"Like this river birch," said Duhr, pointing to a mature birch along the road. "If it starts on fire and it's on this break and the winds are from the south, it needs to go because it would drop outside the unit and start a fire. An ember from thirty-five feet up with a good wind goes a long way. So, we have to cut down trees close to the break that are chimneyed or laddered or smoldering. We prefer not to do it during the burn. We try to wait until the end when we can take our time and make sure it goes the right way. But a lot of times these are edge trees that grow toward the sun, so they have a heavy lean and they do not always go where you want them to go. If we have to, we'll cut them down during the burn and process them later."

Laddering refers to a fuel ladder, which is vegetation near the base of a tree that allows a fire to climb—like the first few steps on a ladder—up the trunk of a tree. The fuel could be brush, vines, dead branches leaning against the trunk, or even an old fencepost. The chimney effect is when a tree burns from the inside; it can send sparks into the wind. On dead ash and oak killed from oak wilt, the bark will sometimes delaminate from the trunk and the fire can get under the bark of the tree. When the sapwood gets punky, it smolders perfectly, so the fire is burning outside of the trunk but underneath the bark.

Duhr continued his story: "We were putting out some smokers. It was an ash. These dead ash have all kinds of weird internal stresses. I was watching him [the sawyer], and he had too fat of a hinge. It's tough to identify what is your good holding wood." The sawyer misjudged the lean, and as the bar got pinched during the back cut, the holding wood snapped off, releasing the Stihl 361. The sawyer stepped away safely, but the tree landed on the saw and crunched it. It was unsalvageable and would now be used for spare parts.

As we ended our tour through the foundation's land, Duhr became reflective as he viewed a blue semi and log trailer parked on a nearby landing. "This is the best-case scenario," he told me. Beside the vehicles were multiple piles of mixed hardwood and softwood logs stacked as high as a house and stretching out to the length of a football field or more. "Like, really? Somebody is going to pay us something and they are going to take it? They are going to push it towards savanna and push it towards oak woodlands. They take the wood and utilize the wood. Now, all we have to do is to go back and do some hand cutting and do some fire break work and do some chainsawing of downed material."

The logging company had harvested most of the logs with a dangle-head processor, but they'd also felled some of the trees with chainsaws. If the logger had decided not to bid on the job, Duhr and his crew would have done it all themselves with chainsaws. He estimated it would take ten years or so to cut that amount by hand.

Of that immense pile of logs, only a fraction was cut by a sawyer wielding a chainsaw. The rest were felled, limbed, and bucked by an operator sitting in a cab. It occurred to me that the chainsaw, a machine that revolutionized the world of logging, was only revolutionary for a few decades until these machines appeared in the woods.

8

The Last Lumberjacks

The Saw in the Hands of a Few

The quest to harvest timber more efficiently did not stop with the development of the modern chainsaw. Eventually the entire process became mechanized until much of the work could be done from the comfort of a heated or air-conditioned cab. The risk-taking, physicality, skill, and woods wisdom of a chainsaw-using lumberjack lost their value in the face of this new technology. More trees were harvested by heavy machinery, and thus, there was less demand for such an occupation. The heavy machinery invented for timber harvesting emerged in the 1970s, just as the chainsaw was becoming lighter, safer, and easier to use.

In fully mechanized logging systems, there are two main machines that get the tree on the ground. One is the timber harvester (also called a feller-processor), and the other is the feller-buncher. The machines can either be tracked or on wheels. The harvester fells, delimbs, bucks, and tops the tree, cuts it off at the stump, and then places the log wherever it is desired. I've worked with one. During my time doing line-clearance tree trimming, my team and I were once dispatched to an area near Oconto Falls where a new line was set to dogleg into a forest. We needed to remove so many trees that it made sense to rent one of these machines rather than have a crew of sawyers do the hand cutting. But I still ended up cutting many of the trees. The harvester often leaves a sizable portion of a tree's stump—especially on uneven terrain—and our practice was to leave stumps flush with the ground. I spent days flushing stumps on inclines and cutting close to the earth. I sharpened my chain often

This swing-boom feller buncher, with the cutting head mounted on a boom, allows the operator to harvest trees from one spot. A drive-to-tree feller buncher, with the cutting head mounted directly on the vehicle, forces the operator to drive up to each tree to cut it. WISCONSIN DNR

throughout the day, bracing the saw against a stump while kneeling on the hillside.

The feller-buncher is even less refined than the timber harvester. The operator drives up to a tree, extends a boom out to the base, and, using the clawlike grapple of the feller-buncher head, grabs the tree and cuts it off at the stump. The entire tree is picked up and "bunched" together with other trees. They can be sorted by size or how they will be used. Then, either a sawyer can wade into the downed canopy and limb the trees by hand or another large machine, a delimber, can do the job. Imagine grabbing a weed, gripping the stem tightly, and running your hand down the stem until all of the leaves are gone and bunched in your hand. The delimber works in the same way.

With the trees on the ground, other machinery can move in. The grapple skidder, with its huge crab claw, can move whole trees around, and the cable skidder uses winches and cables to drag trees where needed. The trucklike forwarder grapples the logs, loads them into its bed, and moves them to the landing—the staging area where logs are piled for transport. A wood chipper might also be present—not the backyard kind used in the film *Fargo*, but an industrial one that can swallow large branches and spit out wood chips with such speed and destruction that it is both amazing and horrifying to watch. Eventually, a logging truck will show up at the landing to haul it all away.

To learn just how much chainsaws are used in modern-day logging, I visited the University of Wisconsin–Stevens Point. I found a parking spot four blocks off campus and hiked to the Trainer Natural Resources Building, an academic hall that has a giant tiled mural on its south-facing wall featuring the Vitruvian Man next to a muskellunge. I sat down with forestry professor Michael Demchik in a conference room, and we talked about chainsaws.

He was the right person to answer my questions. By his count, he had trained 1,364 people on the proper and safe use of chainsaws. The forestry program at UW–Stevens Point is one of the best in the country, and its graduates are trained on chainsaws, brush saws, and wood chippers. They are also S-212 Wildland Fire Chain Saw certified, a key qualification for federal work.

Demchik leads a two-day class that is about a day and a half of inside lecture and a half day of dissecting saws. Students tear apart Stihls, Husqvarnas, and electric saws from Milwaukee Tools and learn how to maintain them. After that, they spend two days felling and bucking trees. They get "stump cards," like report cards that grade their cuts. Demchik observes each student-sawyer as they fell a tree and then assess everything once it's on the ground: the stump, the size of the tree, the species, the tree's conditions, the face cut, the back cut, hinge wood, the direction of the fall, the felling zone, and so on.

"Over 70 percent of the volume in Wisconsin that is cut is fully mechanized. It's the majority of timber," Demchik said. "The people that are hand cutting now are generally merchandising high-quality hardwood logs. There's not very many hand cutters left compared to what there used to be. When you look at the average age of loggers in Wisconsin, we're losing a lot of loggers through retirements, but relatively few people are coming into it. We're losing more than we're gaining."

Merchandising is the process of cutting tree-length stems into the most optimal length for the highest profit. An operator in the cab of a processor might not notice a tree's best cut. The operator rides high off the ground and must look through a windshield, so they see only part of the picture. Hand cutters have the ability to capture more value from a tree. They can look at the tree differently because they are physically on the ground right next to it. They can notice the tree's defects and make their cuts accordingly.

"In one logger's case," Demchik explained, "the impact of his merchandising decisions was about $135 more per one thousand board feet delivered. With the decisions he was making, he was able to capture what would have been great two-saw logs [logs destined to be cut into dimensional lumber], and he was able to make veneer cuts [wood sheets] out of them. So, the impact of someone who is able to do both a good job of running a saw and also see the high-quality pieces in there—they do a better job of merchandising. That doesn't mean you can't with a processor. It's easier for someone running a saw because they are physically able to look at the trees. It's just a different game."

Seeing defects in wood is important for the safety of the sawyer, but it can also prevent damage to the blades at the lumber mill. As part of his lessons, Demchik brings in weird metal objects he's found in trees: spikes, insulators, lots of nails. On one tree he worked on, he found over fifty nails pounded into the trunk. He is still perplexed about that one. His best guess is that some kid had an urge to pound a lot of nails into a tree for fun. His strangest find was a driveway reflector completely enveloped by the wood of the tree, as if the tree had swallowed it.

At some mills, hand cutters go through a stand of trees first and cut down the individual high-value stems. Then, a mechanical processor harvests the rest of the lower-value trees in the stand. The upfront costs of being an independent hand cutter are low (the cost of chainsaws) compared to the price tag of a mechanical processor, but the insurance costs are extremely high because it is dangerous work. Due to the safety issues and the worker's comp costs, some logging operations opt to forgo employing a hand cutter altogether. This void is filled by independent loggers who work outside of the requirements of worker's comp. They work as contractors and are usually paid per one thousand board feet delivered to a landing. The mills often pay them to go in first and fell and merchandise the high-value trees. Alternately, if a stand holds a lot of pulpwood, a mill may send a mechanical processor in first to clear the area and make it easier for the hand cutters to access the spot.

"You may have some sites where the trees are beyond the ability of the machine to cut," Demchik explained. "So let's say 90 percent of it gets cut with a processor but then there are 10 percent of these ginormous trees that need to go out. Those may be cut by hand cutters because they are outside the capability of the machine. They are big or odd-shaped or there is something going on that makes it problematic. In Wisconsin, a lot of it is flat but when you get into the Driftless [Area], it's not. In portions of that, a lot of it is hand cut because of the inability to get the machines in there. Inability is the wrong way to say it—we have the capability to operate processors on very steep slopes with tethered equipment, but I don't know of any of them being used in Wisconsin."

Tethered logging is a rigging system where all of the heavy machines—forwarders, harvesters, feller-bunchers, and skidders—are tied to cables and winches and anchored on top of a hill. This allows the machines to operate on steep slopes that normally would be dangerous. Some loggers in the Pacific Northwest work via tethered machines on mountainsides so steep they resemble black diamond ski slopes. Wisconsin isn't mountainous enough to justify such an operation. If an operation deems a group of trees on a steep slope valuable enough, they'll send hand cutters in.

Demchik cited a workshop the forestry program recently hosted about small-parcel management. "The nature of the land ownership is different" these days, he explained. "The parcel size is going down. People are buying smaller parcels and that is partly due to the cost of land [being] so much higher. It also makes logging more challenging." Logging operations are difficult to carry out on small parcels. Now that the standard logging equipment is large and expensive to operate, logging small parcels is often not worthwhile unless multiple landowners in the same area band together to have their shared forest logged.

"Then there is a public perception issue," Demchik continued. "If you were to look at the general media portrayal of [logging], it's not a favorable portrayal of the industry. This is one of the most renewable and most sustainable industries that can be carried out in the world, and Wisconsin is one of your best in the nation for doing this sustainably."

As I hiked across campus back to my parking spot, I considered pulpwood, which is used to make paper for books, among other things. Usually, pulpwood is mechanically harvested, but in some situations, it may be hand cut by loggers with a chainsaw who can access hard-to-reach spots. Therefore, there is a chance that a chainsaw may have been used in the creation of this book.

Nowadays, hand cutters are a dying breed—the last of the true lumberjacks who do things the old way. But there are a few who take this even further. Wendell Berry is a fifth-generation native of north central Kentucky. He has published novels and books of poetry, but he is best known for his essays about the dignity of work, rural communities, and people's connection to the land. He is also a critic of environmental destruction, gross consumerism, and soul-sucking industrialization in general. After a career in academia, he moved to a farm along the Kentucky River where he uses horses to work his land.

In his 2003 essay "Feminism, the Body, and the Machine," Berry wrote about his marriage and the interplay between the human body, technology, and machines, including his decision not to use a computer. He tried to explain where he draws the line when allowing certain technologies into his life. "I still cut my wood with a chainsaw," he wrote, "which has nothing to recommend it but speed, and has all the faults of an airplane, except it does not fly." He concluded the essay with a vow to seek to escape dependence on machines until the end of his days: "And I am not without hope. I knew a man who, in the age of chainsaws, went right on cutting his wood with a handsaw and an axe. He was a healthier and a saner man than I am." *The New Yorker* published a profile of Berry in 2019 featuring a photo of him next to his well-stocked woodpile. Evidently, he still uses his chainsaw.

Berry has also written about the concept of appropriate technology: using the right tool for the right job. *Appropriate technology* is an umbrella term describing how technology should be wisely applied in order to be environmentally sustainable, decentralized, local, small scale, and economical. The downside of these "appropriate" types of technology is that they are often labor intensive. "Those of us who use draft animals for work on the farm or ranch or in the woods usually know that we are odd," wrote Berry in his essay "Taking Draft Animals Seriously." "We don't fit well into the modern pattern or live up to the modern ideal, which applies relentlessly to every task the maximum available power with the greatest possible speed."

To gain a better understanding of these "odd" individuals, I spoke with Taylor Johnson, a horse logger in Hayward, Wisconsin. This is a specialty form of logging where the sawyer fells, limbs, and bucks trees with a chainsaw, then uses horses to skid the logs out to the landing. Horses can navigate woods with little disturbance. They are also quiet and more environmentally friendly than mechanized logging operations. Horse logging is often appealing to landowners of smaller parcels and those who cringe at the war zone–like havoc that machines can wreak in woodlands. The trade-off is lower productivity and less money for the landowners, since horse loggers cannot match the massive amount of work mechanical crews can quickly complete. However, with the nimbleness of the horses and the boots-on-the-ground perspective of the horse logger, timber can be targeted and selectively cut as part of a larger forest stewardship plan. A one-time clear-cut with big machines is usually a quick-money transaction, while a horse logger harvesting small amounts of timber over many years on a property is a long-term investment. This is the kind of appropriate technology that Berry endorses. "Put to appropriate tasks that are appropriately scaled, [horses'] work gives a favorable practical result and a favorable economic return," he wrote. "If that were not true, the draft breeds and the teamsters' skills would not have survived."

Johnson started logging in the eighth grade. His first job was running a measure stick and bunching up wood from September until breakup—the time in late winter or early spring when the frost comes out of the ground, roads are closed, and loggers cannot truck wood. Some sawyers take time off work during this season, but not the crew that employed Johnson. The trick is to have a pine job lined up. The crew moved to a pine plantation near Trego where the sandy ground could not turn into mud. Johnson grew up logging with his father, but this, as he described it, was his first "Here's your saw, now go to work" job.

His father kept horses and mules when Johnson was growing up, but then their family logging operation went mechanized. "We had about twenty-some guys sawing for us, and my dad hated it and I hated it," Johnson recalled. "Everything started to go mechanized—this is back in the late nineties—but in about '96 we started talking about getting back to horses." It wasn't until years later, after logging with his father and on other crews and doing a two-year stint as a house builder, that he formed his horse-logging operation, Skinner Horse Logging.

Johnson was fifty years old when I spoke with him. The other hand cutters he knows are older than he is. Eventually, they will retire, and then who will replace these men? The last of the lumberjacks who are already few in number? Whoever it is, that person will be in demand. "For any young man that wants to get into running chainsaws," Johnson told me, "I think there's going to be more money in it in the next five to ten years than there's been in it for the last forty. Even just running chainsaws and felling trees for these processors and cutting quality logs. When you're cutting that log, sometimes a decision on two to three inches in length makes a difference if that's a saw log or a stick of veneer depending how much sweep [curvature] is in the log. If I'm looking at a tree and it's all stretched out, I'll get down on the ground sometimes to see where the sweep is. If you are sitting in a machine and looking at one side of it, looking through a window from ten feet away, you're not getting everything you can get out of these trees."

He continued, "One of the things that changed the most since I started was there [used to be] big crews of guys working. You got to a job and there's three to five guys running saws. You go to the saw shop in the morning, and there's a bunch of guys there having a cup of coffee and Jack would put out cookies." He was referring to Robert Jack Kelley, owner of Lumberjack Sales and Service in Hayward and one of the loggers who blazed the American Birkebeiner Ski Trail in the 1960s. "Loggers are always very competitive. Who can cut the most? Who is the most accurate? Especially amongst us younger guys—we were always trying to see who could fell a tree the most accurately. That's gone. I imagine there are places where there is more than one guy working, but a lot of times the guy that is working is in a processor. There are very few young guys doing this."

The breakup season doesn't affect Johnson much anymore. With his horses, he skids out logs early in the morning before the ground starts to thaw, and then he spends the afternoon felling and cutting. The same goes for the hot days of summer. He's out there at daylight, skidding logs in the cool morning. As the sun comes out, he hitches his horses up in a shady

Taylor Johnson and his horses tread more lightly upon the soil than logging machinery, and they are also able to navigate through woodlands and extract logs on narrow trails. PHOTOGRAPH © CARL COREY

spot outside of the felling zone and resumes cutting logs to be hauled out the next morning. The only time of the year when he might take a break from the woods is between late summer and the first frost. "You don't want to be hitched up to a big log with a team of horses and run over a ground hornet's nest," he assured me. "They get more and more aggressive all summer long. You start getting those cool nights in August and September and it warms up in the day—it's like they're half nuts." Johnson has been stung and so have his horses.

Johnson works with Belgian Draught horses that are about five feet tall at the withers, as well as smaller Norwegian fjord horses. About the Norwegian breed, he said, "They are a smaller horse, but pound for pound they are so strong it's crazy. One of the mares chased a three-hundred-pound bear through the pasture. I can feed almost three of them for [what] it costs me to feed one Belgian. They are so efficient because they come from these desolate mountains in Norway where there's not much to eat. If you look at the old cave paintings, that's kind of how they look. They have big bodies, big necks, and short little legs. That's an exaggerated version of what these are. They have that big arched neck, and their manes grow straight up, and you've got to keep them trimmed. It almost looks like a Mohawk, and they have a black stripe down the middle."

Johnson uses a logging arch—a wheeled tool for moving logs—that looks like a chariot. It lifts the butt end of the log off the ground when the horses move forward so that just a few inches of the log drags on the ground. This type of logging arch adds a mechanical advantage to the horse-powered process of moving heavy logs, and it allows the horses to work at a natural pace. Johnson usually has two hitched up, but he sometimes harnesses three abreast if the terrain is rough. Other times, he uses one horse when moving smaller loads. He transports his horses to the jobsite every day. "If they are acting really weird and scared, I know that I will see wolf tracks on my jobsite within a day or two," he said. "I used to leave horses on the job overnight and set up a corral. I don't anymore because that's asking for trouble. We have a ton of wolves here. When I'm working with young horses, I always take them into the woods with me so they can watch and see what I'm doing. It's all about developing a bond with that animal. The horse trusts me before I head into the woods."

Years ago, Johnson pulled his team in a horse trailer, but now he transports them in a refurbished U-Haul box truck. He subdivided the front area into a mini workshop where he fixes saws and hangs harnesses. Working as he does in remote areas that are often thirty miles from the nearest town, he needs to make repairs in the field. He also brings grain, hay, and water for the horses. Many of Johnson's customers have waterfront property that can be accessed only by narrow winding roads—poor places for turning around with a trailer. The box truck—like his horses—is far more nimble than the alternative.

Landowners will hire Johnson and pay him by the cord or foot. First, he lines up a log buyer, then the log buyer writes the landowner a check, and finally the landowner writes a check to Johnson. He has long-term relationships with almost all of his customers, who have paid him to selectively cut their land a little at a time over years. One property has been through three different owners during the time Johnson has worked there—each owner has recommended Johnson to the next one. Some of his customers are wealthy; others are more middle class, like the retired schoolteacher with a small woodlot. Many see Johnson and his low-impact operation as part of their stewardship of their forestland. "It's so low impact its crazy," said Johnson. "I've done some work for the State of Minnesota, cutting walnut in Cannon Falls. In this area were these endangered plants called trout lilies. They didn't [want] big machines in there because they didn't want anyone messing up those plants. We cut the walnut out of it and never hurt anything. We pulled walnut out of there that was twenty foot long with these carts and a team of horses through an area that was not fifty-four inches wide. Our impact is real minimal."

In other arrangements, Johnson works with landowners on a timber-stand improvement program, also known as a worst-first forestry program, where he removes low-value trees first and leaves the higher-value trees to grow and become more valuable. Johnson described an instance when he visited a potential jobsite where a stand of young maples was marked for removal. He asked why the trees were marked. The landowner explained that another company had marked them to come down because they were blocking the path of the machines. Johnson said he could log the entire area with his horses and leave the young maples untouched simply by cutting out one ironwood tree that was in the way. "I can cut one truckload of wood and it's profitable for me," he added. "Most of these guys [mechanized loggers], to cover their moving costs and time, they need to cut twenty or thirty truckloads of wood just to show up. [And for the landowners,] it doesn't pay to [have them] go in there and wipe the woods out and leave it like a mess for the little bit of money you get out of it. Those trees are worth more standing if you're going to sell [the property] in the next fifty years."

Johnson had fond memories of participating in an event called Log a Load for Kids. It's an annual fundraiser where loggers donate the value of a load of logs to their local Children's Miracle Network hospital. Logging operations work together to log a property, and the proceeds go to the charity. Schoolchildren are bused out on a field trip to see the logging in action. Johnson was the only horse logger there. He was also the most

This two-horse team is hitched to a go-devil, a sled used to skid individual logs through narrow passages in the woods. The front end of the log is placed in the go-devil to keep the log from digging into the ground. WHI IMAGE ID 58515

popular, as the kids gravitated to his horses. At lunchtime, another logger baited Johnson by saying that his horses do far greater damage to the forest than machines. Johnson already knew what the man was referencing—the pounds-per-square-inch theory.

"If that's the case, you know what's really tearing up these woods really bad?" Johnson countered. "It's these big whitetail bucks that run and jump and land on those little pointy feet. They're really tearing things up." Everybody laughed, including the other logger, but he continued to argue that horse hooves were harder on the land than his machine. This particular logger's machine had four tires on the front and four on the back with a track stretched between them. Johnson estimated the machine must have weighed over fifty thousand pounds.

The pounds-per-square-inch theory was addressed in a study by Joshua A. Deal and summarized in a paper titled, "Site Disturbance and Social Impacts of Animal Logging and Conventional Logging in

Appalachia." The study analyzed the work of two loggers. The first was Bill Best, who did a silviculture clear-cut on a section on Virginia Tech's Fishburn Forest using a tracked shear, skidders, a delimber, and a forwarder. The other logger, Jason Rutledge, performed single-tree selection logging in 2002 on a site in Copper Hill, Virginia, using a chainsaw and draft horses pulling a logging arch to a loader mounted on a truck.

Deal used a point-sample method by recording soil disturbances across the two sites and ranking them on a scale of one to eleven. The results showed that the horse-logged site had the most undisturbed area at 56 percent, while the undisturbed area of the mechanical operation was at 20 percent. One percent of the horse-logged land had what was called "deep disturbance," or disturbances greater than five centimeters deep, as opposed to the machine-logged area, in which 11 percent of the land had deep disturbances. The abstract concluded: "Animal logging offers less soil disturbance and has higher post-harvest aesthetics than conventional logging. This is primarily due to the fact that animal loggers perform single tree selection harvesting while conventional loggers perform commercial clearcuts. However, animal loggers can economically harvest timber using singletree selection and inflict less damage to the residual stand."

The majority of Deal's observations point out the benefits of low-impact horse logging. However, there was one paragraph in the study that the machine logger had latched onto and used to taunt Johnson: "An animal can compact the soil just as bad as a skidder on active areas such as landings. This is due to the high number of pounds per square inch exerted by the animal's hooves. Due to the weight of logs and equipment, severe compaction is inevitable on harvested areas. However, the use of animal extraction limits these areas on a logging job."

Johnson explained it this way: "When [the conventional loggers] drive, those tracks or wheels don't stop when they hit a little sapling. It's a continuous compaction. I know they go by that theory, but the grousers on those tracks sink in farther than my horses' feet ever do, so I don't even believe it. But let's give them the benefit of the doubt and say that it's true. When my horses hit a one-inch sapling with their foot, they move their foot around it just like we would. They feel that and step around it. Let's say their foot sinks into the ground. Right next to it, there's still grass and brush and trees that are not affected. So those roots keep growing and break that ground up. With the tracked machines, you go there in the spring when the snow comes off; you can look at those tracks for the next twenty years."

As a logger, Johnson has taken continuing education courses, but half of the material doesn't apply to him—things like erosion control and bulldozing grades and handling chemical spills. He carries just a few gallons of gasoline and bar oil into the woods—a minuscule amount compared to the toxic liability of hundreds of gallons of hydraulic oil or diesel that conventional logging machines could leak onto a jobsite.

Johnson once participated in a horse-logging study. The Minnesota Department of Natural Resources selected two test sites in Rochester in a wealthy neighborhood overgrown with buckthorn. On one site, the buckthorn was removed by cutting and treating the stumps with herbicide. On the other site, Johnson and a partner cut the buckthorn at about chest height, wrapped a chain around the stem, and pulled it out by the roots with a single horse. They skidded out the tops and root systems in bundles. Five years later, he was told that, dollar for dollar, horse-powered removal of buckthorn was the better option.

Johnson cuts with a few Husqvarna 375s and one Stihl MS-261. "It isn't always whatever the best saw is," he stated, "it's where you get the best service. If you take Husqvarna, Jonsered, and Stihl, they were a horse apiece for years on which was the better saw. If Jack [Kelley, in Hayward,] was working on a lawnmower for somebody, and I walked in with my chainsaw, he would drop everything he's doing. He catered to loggers. He used to be a logger and knew how damn hard it is. He [once] had a guy speak up to him, and [Jack] said, 'Listen, this guy makes his living with this and you're mowing the yard. He's got to get back to work, so maybe you want to go eat lunch or something but I'm fixing his saw now.'"

The solitary nature of Johnson's work adds a certain amount of risk to a job that is already taxing. A week earlier, a hanger from a dead pine that was caught up in the branches of another tree fell on top of him. His hard hat went flying, and it knocked the saw out of his hands. Sometimes his son Sullivan works with him, but when he's alone, he has a system with his wife, Annie, to call or send her a text message. If she notices a lapse in communication, she will know something went wrong.

"I don't think even your average concrete guy or construction worker has any idea of how hard it is to make a living logging with a chainsaw,"

Johnson speculated. "I figure if I use two tanks' worth of logs out of my saw, I've worked harder physically than if I had worked all day framing a house. Because I'm working by myself, I don't have to cut as many tanks out, but in the past [when I worked on mechanized logging crews], we used to run out seven to ten tanks a day. Every sawyer had to cut anywhere from a truckload to a truckload and a half a day to make a living. Then you had a few rare animals that would cut a couple truckloads a day. We had a guy that, twenty cord a day, that's what he averaged. That's two semitruckloads a day, hand cut. You'll hear guys talk that they cut twenty cords a day—they are telling you about the best day of their life. I've never cut twenty cords in one day. I'm not that fast, but I'm an accurate cutter. I like doing a clean job, so speed was never pushed on me. If you are moving fast enough to cut a truckload a day, that's a hard pace for the average working guy."

A logger's cord has the same dimensions as a cord of firewood. Instead of a pile of firewood, it's a pile of logs. It is four feet high, four feet wide, and eight feet long, adding up to 128 cubic feet. Some of Johnson's higher-quality logs are sold per thousand feet. One thousand feet equals 2.2 logger's cords, so Johnson needs to bust out between one thousand and two thousand feet of logs per day to make decent money. He aims for three-cord days.

To convey the strength and stamina needed to do logging work with a chainsaw, Johnson told a story of a man in his sixties who came from a logging family. He wore old boots and a ripped-up shirt and looked pretty bedraggled. He was about five-foot-ten, weighed in around 160 pounds, and usually had a beer in his hand by nine in the morning. Other loggers in the community called him a bum, but this always steamed Johnson, who had seen the man work on a feller-buncher operation. The man's job was to delimb the trees by hand with a chainsaw. He limbed forty cords in one day by himself. "That guy's an animal," said Johnson. "That guy would work 90 percent of the good men working into the dirt. You grab the average toughest twenty-five-year-old man you can find on a construction site and throw a chainsaw in his hand and tell him to run that many tanks out in the firewood pile—he couldn't do it. That old boy is walking around out in the woods and brush and stumbling and climbing up on logs. He works his guts out. He might be a drunk or something else but don't say he's lazy. And he does it five days a week."

9

Bold Eagle, Cute Bear

The Saw as an Extension of an Artist

After woodcutters deforested the eastern forests of Maine, Pennsylvania, and New York to supply lumber for the growing colonies, they migrated to the immense pineries of what is now Michigan, Wisconsin, and Minnesota. In the Chippewa Valley of Wisconsin Territory, the stands of white pine grew denser than in any other part of America the European immigrants had explored up to that point. The Chippewa River and its upper, fast-water tributaries flowed through these pine forests and the river emptied into the Mississippi, the main artery leading to the markets of the new world.

In the late 1800s, lumberjacks worked the vast unbroken forests. They cut down sugar maples, yellow birch, elm, basswood, red oak, jack pine, fir, balsam, and hemlock, but they focused on one species in particular. The white pines of the Chippewa Valley were nation-building trees: light yet strong and the best wood for all-around construction. They were easy to dry and slow to rot. Some white pines grew to be more than two hundred feet tall and five feet thick and yielded thousands of board feet of lumber.

Logging camps sprang up in this region and eventually came to employ more than twenty-eight thousand men, many of whom were illiterate European immigrants. It was here that apocryphal stories spread of a supernatural lumberjack by the name of Paul Bunyan. The loggers felled the pines by axe and saw and then cut them into sixteen-foot lengths. They stamped them with the logging camp's logo—like branding a steer—and skidded them out of the woods on go-devils, hauled them on ice roads to the nearest river, and floated them downstream to mills during the spring floods.

The epicenter of this logging boom was Eau Claire, Wisconsin. It came to be nicknamed Sawdust City as it was the spot where three separate logging towns eventually merged. In the 1880s, it was believed to have more sawmills than any other place in the world. During this logging era, an estimated six hundred million logs were floated down the Chippewa. These harvesting methods were not sustainable, and by 1920, all of Eau Claire's lumber mills were gone. Fresh-cut sawdust has recently dusted the ground in Eau Claire again, however, and it was produced by chainsaws. Every summer from 2011 to 2024, the city's Paul Bunyan Logging Camp Museum hosted the US Open Chainsaw Sculpture Championship that was founded by Gordon "Wally" Wall in 2010.

People started using chainsaws as carving tools in the 1950s as one-man saws became smaller and more maneuverable. The chainsaw created an entirely new American folk art, practically overnight. Few other modern handheld tools have had such an effect on the world of art, except for maybe aerosol spray paint (for urban murals) and arc welding (for metal sculptures). It was easy for people to experiment with chainsaw carving because wood is cheap and if a sculpture didn't turn out, it could be tossed into the woodstove. Chainsaws are also multipurpose tools, and as they made their way from forests to farms to backyards, more people had access to the tool for artistic experimentation.

The role of the chainsaw in the world of art is exactly the same as its role in the world of logging: it is a labor-saving device. It saved sawyers from back-breaking work with the crosscut saw, and it saves wood sculptors from the labor of chipping, chiseling, and cutting away at a large block of wood with hand tools. The chainsaw increases productivity for both the artist and the lumberjack. A skilled sawyer with a chainsaw can fell a tree with a trunk as big around as a basketball hoop in less than a minute. A skilled chainsaw artist can carve a small sculpture so fast that the activity can be considered performance art. Watching a sculptor do this with hand tools would put you to sleep, but you can pull up a lawn chair, crack open a beer, and watch someone with a chainsaw very quickly turn a log into a cute bear cub.

This particular niche of the folk art movement grew as the chainsaw evolved into a lighter, smaller, and fast-handling power tool. Early chainsaw artists in the 1970s formed guilds, wrote newsletters, and published how-to books on the subject. The art form also got a big boost in the 1980s with carving exhibitions at the Lumberjack World Championships held in Hayward, Wisconsin. Chainsaw carving get-togethers and exhibitions spread across the country. Of course, the internet has also advanced the art form, allowing artists to showcase their work in online galleries, accept commissions, and study the best practices of their craft. Like other forms of folk art, chainsaw carving has received its fair share of criticism within the art world, but I believe it's unwarranted. What I appreciate is that the art form is egalitarian, and the skill cannot be faked. The mastery is right there for all to see, carved in wood.

I heard the 2023 US Open Chainsaw Sculpture Championship before I saw it. From the parking lot, my ears pricked up at that high-speed, *yak-yak* two-stroke sound, an orchestra of saws operating at once. Foam ear plugs were being handed out at the ticket table. This annual event is held at Eau Claire's Carson Park, a 134-acre multiuse green space. The park is a peninsula surrounded by an oxbow lake that was once part of the Chippewa River—that one-time highway for millions of logs floating to the city's mills. The scene felt like a small-town festival, complete with food vendors, displays by local woodworking clubs, and tents featuring chainsaw artists not competing in the championship. The big show was under a giant carnival tent, and the sides were sealed off by mesh netting and pedestrian barriers. This was for obvious safety reasons, since look-over-your-shoulder gawkers and chainsaw teeth do not mix. I arrived on the last day of the three-day event. Some artists were nearly finished and painting their work, while others were still making detail cuts on bare wood.

The artists used all kinds of chainsaws, though small ones with pointed carving bars were the most common. At the individual carving stations, each artist had a small collection of saws, along with many other tools, including angle grinders, sanders, and cordless drills with various bits. Extension cords snaked everywhere. The artists used air compressors and leaf blowers to blow the sawdust off their sculptures and snow shovels to scoop it off the floor once it got ankle deep. Gas cans, paint cans, tool chests, and scaffolding lay strewn about. Some artists used blowtorches to add a burned effect. And log stumps served as handy utility tables. I

A guest carver at the US Open Chainsaw Sculpture Championship in Eau Claire, Lisa Doeren of Renegade Carving Company in Abrams, Wisconsin, uses a saw with a pointed carving bar.

circled the huge tent many times, watching the artists and taking note of their names, nationalities, and local sponsors, all of which was posted on the mesh netting behind them.

The sculptures included an action scene of a fox lunging at a flushed ringneck rooster, a futuristic standing bear that looked like a Transformer, what appeared to be the pagan mythological figure of the Green Man adorned with tree frogs, a skeleton against a crucifix-shaped deadwood tree, and an American Indian chief wearing war paint and a full headdress with a bald eagle perched on his arm. Wisconsinite Mike Koeppel carved a driftwood stump with a muskie, the state fish, on one side and a snapping

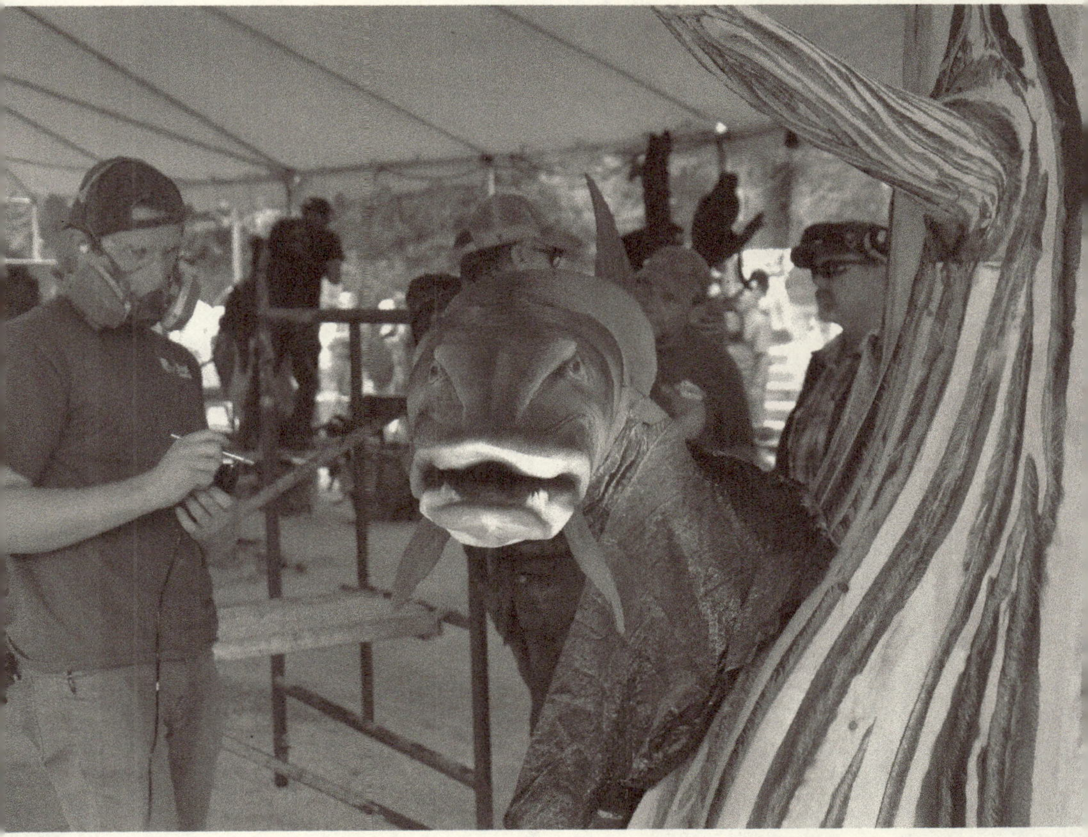

Mike Koeppel paints the finishing touches on his muskellunge at the 2023 US Open. He wears a respirator for the paint fumes, ear plugs for the constant chainsaw noise, and safety glasses for sawdust and flying wood chips.

turtle on the other. John Hayes of Ireland—who is known for carving Dragonslayer, the longest sculpture cut from a single piece of wood—made a collage of swans.

Out of all of the carvings, I thought the one made by Takao Hayashi of Japan was the most visionary, creative, and strange. It was his tenth US Open appearance, and he had won it four times. I could tell he was a perfectionist because while almost all of the other carvers were painting and adding final touches, Hayashi was still carving his masterpiece. It was the largest sculpture of the group: a vulture perched on the shoulders of a squatting monkey balanced on the open jaws of a crocodile emerging from

Ankle deep in sawdust, Takao Hayashi did not have time to paint his sculpture during the allotted time at the US Open championship. However, he drew the most spectators as he continued to refine his masterpiece with small, detailed cuts.

a lotus. As far as I could tell, the amazing sculpture looked complete, but Hayashi kept making tiny detail cuts here and there with a pointy Samurai Legend carving bar. Next to him, his fellow countryman, Hikaru Kodama from Hokkaido, worked on a mermaid riding side-saddle on a seahorse. Kodama is considered one of the best carvers in the world for human figures, but by pure chance, the wood that he was using that day revealed a knot on the mermaid's left cheek, leaving her with an odd birthmark.

As each artist finished, an army of volunteers would move in and wheel their sculpture into a line. A crowd slowly assembled in the bleachers after the judges reviewed the final pieces. The announcer explained to the audience that the judges had named a first-, second-, third-, and fourth-place finisher, with everyone else sharing fifth place. "I think they all should be first place," he said, "but they would still try to figure out a way or pick at each other to see who's first place." Justin Driver of Kentucky took fourth place with his skeleton carving. Chris Kuehn of Minnesota came in third with his fox flushing a pheasant. Chris Woods, from Wales, took second with his pike slashing at a duck, and Canadian Jake Rhodes won first place with his bull moose mount. Next, the carvings made during the championship were auctioned off, along with numerous other carvings made by guest carvers over the weekend. Rhodes's winning moose sculpture sold for $7,200.

After summer passed, I was driving around town and noticed a totem pole–like carving that had not previously existed in a yard I drove past regularly. It seemed to have appeared out of nowhere. The owner had pruned the tree and left the spar for sculpting. Parked near this was an enclosed trailer with the name Buschke Carving on the side. I was familiar with the name. Ben Buschke was a local chainsaw carver I'd seen carving a few times at the Waupaca County Fair. I contacted him and arranged to visit him on a cold October day at his workshop just outside the small town of Manawa, Wisconsin. When I pulled up to his house, I found a carving of an eagle with flared wings sitting in his driveway. It was loaded on a pallet and ready to be picked up by a freight company. Inside his garage, I discovered more finished carvings ready to be shipped, including a life-size Smokey Bear figure.

When I mentioned that I'd visited the US Open in Eau Claire earlier in the summer, he said, "I've contacted them and got on a waiting list. There's one guy I carve with in Scandinavia [a nearby town] every year, and in the last two years he got into it. Michael Koeppel. He's really good." I agreed, remembering Koeppel's amazing muskie and turtle sculpture at that year's open.

We hiked around the side of Buschke's house to the carving shed, which was furnished with a workbench, a tool chest, whitetail skulls and antlers on the wall, and a woodstove blazing in the corner. The stove was made from a metal barrel; as he threw in a slab of wood, Buschke told me he built it from a kit. He cracked the garage door to balance out the heat kicked out

Ben Buschke temporarily stores his chainsaw creations in his Manawa garage before shipping them across the nation. His sculpture commissions have ranged from Smokey Bear and SpongeBob to a re-creation of a client's beloved Labrador retriever.

by the woodstove. Sawdust and wood chips covered the floor. His current project was an owl poking its head out of a tree crevice. The wood block sat on a worktable that could be adjusted up and down. Under the table were sheets of plywood full of cuts and gouges covering the concrete floor. A cutting board was positioned under the owl sculpture. The plywood was there to protect his chain from the concrete and the metal frame of the table.

Ben Buschke grew up on a dairy farm where chainsaws were used regularly. His grandfather had Stihl saws and used them for cutting firewood as well as removing fallen trees from his fields. Buschke's first idea to carve wood with a chainsaw wasn't from inspiration but from indifference.

"[In 2017, my wife and I] were at the campground at Bear Lake," he recalled, "and we were leaving with our kids. There is a carved bear in the parking lot. I walked by it and looked at it and thought, 'I think I can do better than that,' and I had never done it before. That put the idea in my head." He didn't want to borrow his grandfather's nice Stihl saws for his first foray into carving, so he bought a cheaper Poulan saw. He had no idea what he was doing. As a child he liked to draw, but that was the extent of his artistic experience and training. In 2016, he carved a four-foot-tall bear over the course of a few days. He still has it in his basement and describes it as "pretty rough looking."

He doesn't remember what kind of wood it was, but he knows it was a dead hardwood. It was a poor choice for a beginner. If he could do it over again, he reflected, he would start with a green softwood log and have a sharper chain. It's harder to cut deadwood than it is to cut green wood. There is less moisture. But on the other hand, a sculpture carved from green wood still has to dry out, and this can create the splits and cracks known as checking.

"My second [sculpture] was a little eagle. My wife, Lindsay, said it looked like a turkey, but I thought it looked good," said Buschke as he laughed at her remark. "I've still got that one, too. Every time I did it, I learned a little more. Even stupid things that are obvious like keeping your chain sharp. You realize the difference it makes." He pulled out a Stihl 2-in-1 filing guide. "I like these because it also takes the rakers down a little at the same time."

After the bear and the eagle/turkey, he continued carving. He had no mentor and did not read any how-to books. He watched some videos online but found that many showed less of the real techniques and more of the end results. He gave a few of his early creations away as gifts, and then people started asking him to carve things. Every summer, the town of Scandinavia hosts a corn roast. It's a small-town festival, and one of the events is a chainsaw-carving competition. Buschke thought his work was good enough, so he entered it. Provided with a log and working for the first time under a time limit, he carved a howling wolf wrapped around the base of a tree with an owl on an upper branch. Buschke came in fourth out of nine carvers, some of whom were out-of-towners. The winning sculptures were ranked by the price that the carving was auctioned for. It was the biggest carving he had ever made. "That gave me some confidence," he said.

Between farming and his growing children, Buschke had little time to carve as a young father. So, he made a radical lifestyle change to further his art. He got a job working nights on the maintenance shift at the Waupaca

Foundry, which allowed him to carve during the day. In 2019, he started thinking about chainsaw carving as a full-time vocation. "I felt like I was getting good at it," he explained. "I was starting to get orders. I had it in my head to work there at night, do this during the day, save up money to get more saws and tools and eventually a trailer, and once I felt like I was going good enough and built up some clients, I decided to go for it."

He created a logo featuring a chainsaw, which he emblazoned on his enclosed trailer, and ordered "Buschke Carving" baseball caps. He built a website showcasing his work and printed business cards, and his wife came up with the idea to provide customers with a printed card with instructions on how to care for wood sculptures as they age. He mentioned that some people like the natural cracking and checking that occurs as wood ages, but some people freak out. "I always tell people you want them to dry as slow as possible," he said. "Like if I put them in the garage and out of the sun where it's generally cooler, they don't crack as much. I also put relief cuts in the back. It helps give the moisture somewhere to go. Like on a bear, I make a vertical relief cut down the back since nobody is going to look at the back of it anyway. It's amazing to see sometimes as the relief cut will flex. You will notice over a day or two."

With the purchase of his trailer, Buschke became a mobile on-site carver and an entrepreneur who could showcase his work at events. He built a work bench in the trailer, along with a space-saving storage area for his saw bars and powerheads. Of the ten chainsaws of various sizes that he owns, he won several in quick-carve competitions. At these events, the artists have one hour to create something, and the carver whose sculpture gets the highest bid often wins a new chainsaw.

Buschke usually starts his work with a Stihl 365 with a thirty-six-inch bar for cutting off the biggest slabs to reduce the wood to a rough shape. From there, he reaches for a Stihl 251 with an eighteen-inch bar to cut away smaller pieces of wood to refine the shape. After that, he grabs smaller saws with carving bars. "Those I use the most," he explained. "They are the most versatile. They can cut big chunks, but you can do little stuff, too. They make a big difference."

He discovered carving bars during his first competition in Scandinavia. "It's when you get in and have to cut weird shapes and different angles," he said. "With those bigger, round-nose bars, you can still do it, but it's much harder." He pulled out a Stihl battery-powered saw with a Cannon Carver dime-tip bar. Carving bars are described by the size of the coin that most closely matches the size of the bar tip. The three main types are the dime tip, quarter tip, and toonie tip, named after Canada's two-dollar coin. Since the dime tip has the smallest radius, it is used for the finest of detailed cuts. Also, with these small, pointed tips, the chance of kickback is greatly reduced. The length of the bars used by chainsaw artists ranges from eight inches to two feet, and almost all require specialized chains.

As we talked about carving bars, I noticed that Buschke's Stihl battery-powered saw was sitting out. He told me his preference is to use gas-powered saws, but he said he goes back and forth between gas and electric depending on the season. In the summer, he keeps the door open and allows the exhaust from the gas-powered chainsaw to dissipate. In winter, however, in order to retain the heat from his woodstove, he keeps the door shut and uses the electric saw. "I really don't like anything about the battery saws other than they are quieter and you can carve inside," he said. "Otherwise, I'll do gas all day. It's the power. I like the noise."

When Buschke gets a commission, he sketches out images and he and his client go back and forth on what the final sculpture will look like. He does not draw on the wood, but he knows of many carvers who do using chalk, lumber crayons, or spray paint. Buschke, though, sees the image in his head. He calls it the silhouette.

He gets his logs from a couple of different places, including a local arborist service that is swamped with them and friends and acquaintances who have trees they'd like him to take. His preferred wood is white pine because it is soft. Hardwoods almost double his carving time. Ash is the worst because it quickly dulls a chain. "Oak actually carves pretty well," he added. "You'd think it would be really hard but it's not bad." Before he begins, Buschke checks the wood for nails with a handheld metal detector that looks like a flashlight.

In 2022, he traveled to Pennsylvania to participate in the Ridgway Chainsaw Carving Rendezvous, an event that attracts carvers from around the world. As the premier event of the Appalachian Arts Studio, an art gallery that showcases chainsaw sculptures, the Rendezvous earned the small Allegheny town of Ridgway the nickname Chainsaw Carving Capital of the World. Buschke flipped out his phone and scrolled through a photo album to show me his carving from the Ridgway event. It was a scene from an imagined barroom, and the countertop was presented as the water's surface. On the left end of the bar, the tailfin of a northern pike poked up through the bar top, in the middle were some cattails, and on the right end, a piece of driftwood protruded from the countertop's surface. Under the water, the rest of the pike could be seen chasing two bluegills. The bar was supported by a mass of driftwood. With some leftover time and wood, Buschke carved a couple of stools to round out the ensemble. He easily sold this work.

As a mobile carver, Buschke has created many totem pole–like carvings—usually for people having tree work done in their yards who ask the arborist to leave the spar standing to be turned into a cool creation. He has two levels of scaffolding for these jobs. The largest pole he carved was twenty feet tall and had fourteen different animals carved into it. The client was a hunter who had shot a moose during a trip to Canada. On the top of the pole, Buschke carved a life-size moose head, and the antlers from the moose the hunter killed were mounted onto the wood. He's done some work inside, too, such as carving directly into the walls, siding, and stairways of log homes. At Road America, the historic road-course motorsport racetrack in Elkhart Lake, Buschke has carved stumps around the racetrack into everything from animals to automobiles to antique gas pumps.

In addition to carving bars, another very important tool for any chainsaw artist is a power gouge. This attachment can be affixed to a powerhead in place of a normal bar; a belt replaces the chain and is attached to a pulley. The belt spins a metal wheel about the size of a hockey puck with small notches along the edge, and the tool makes gouges into wood that can range from shallow grooves to deep trenches. When used in one spot, a power gouge can make bowl-shaped depressions or even hollow out a log. A cousin of this device is the debarking tool, a similar attachment that works like a planer—the drum mounts on the saw bar, and when the operator runs it over the surface of a log, it chews off the bark.

Buschke said he liked to use the gouge for making grooves in his sculptures that resemble the striations of tree bark. He fired up his saw and demonstrated the gouge on the sculpture of the owl peeking out of a tree hole. The tool ate through the wood easily, leaving a shallow track. Within a minute or so, he'd made a series of cuts that would easily have taken

By attaching a power gouge to a chainsaw, carvers save many hours that they would otherwise spend carving grooves and depressions into wood by hand.

him an hour with a hand tool. As I watched Buschke gouge the tree bark grooves, I realized that he was a self-made man, a self-trained artist who gambled on a night-shift schedule to practice his craft. His investment paid off. The chainsaw allowed him to carve out a life for himself on his own terms.

Practical Cuts: Starter Projects

After witnessing the carving championship and talking to Buschke, I had an itch to try my hand at chainsaw carving. I did not have a carving bar, but since I was a beginner, I figured I could make do with what I had and just fool around. There was just one small problem: I did not know what I wanted to carve. I did not want to carve a cute bear with a welcome sign to place by my front door. I did not want to carve a vulture perched atop a monkey balancing on crocodile jaws. I wanted to start with something useful and practical and work my way up.

It was bowhunting season, and I was spending many quiet hours in the colored forest watching the leaves fall from the trees. Day by day, the trees revealed themselves. I studied the trunks, their limbs and joints. In my mind's eye, I cut the wood here and there and envisioned what was left behind from the invisible cuts. It was a good form of meditation while waiting for a deer to come along.

Two designs emerged, and both were simple, elegant, and practical. The first one I called the Jim-Cut Chair, and the other I called the Outdoorsman Ottoman. Both of these projects are a perfect mix of work and play, and I'd recommend them to anyone looking to experiment with chainsaw carving. When you find yourself felling and cutting a tree for firewood or milling, just leave a portion of the stump intact to use as your carving medium. But first, you need to make some wedges—another quick and easy project.

A felling wedge is a tool that can lift a tree in the direction it should be felled. It is hammered into the kerf of a back cut pointing toward a notch cut. The wedge also helps keep a chainsaw bar from being pinched when a tree leans on a back cut. I most often use wedges when flushing very large stumps where the weight of the stump sinks down and threatens to pinch the bar near the end of the cut. Wedges can also save the day when a sawyer is bucking a log, misjudges the compression and tension wood, and pinches the bar. A wedge can be pounded into the kerf and create just enough of a gap to release the saw.

Felling wedges can be made of steel, magnesium, or aluminum. But metal wedges are heavy, and nicking one with your saw teeth will damage your chain. Also, wedges have a tendency to pop out when they are not hit squarely, and nobody wants a chunk of pointy metal flying at their face. Plastic wedges made of acrylonitrile butadiene styrene (ABS) are very common, and the leader in that market is K&H out of Medford, Oregon, which introduced its "red head" wedges in 1962. These get their nickname from being red on one end and white on the other. Some wedges are smooth on one side with small triangular spikes on the other to prevent the wedge from slipping out of the wood. Others have rifling or grooves to help keep them in place. Some plastic wedges, however, get brittle in subzero weather and crack apart.

Wood wedges are less durable than metal or plastic ones. After a wood wedge has been pounded with multiple strikes, its wood fibers fray. But this is fine. Homemade wood wedges are free and meant to be expendable. They can be tossed into the kindling box or left in the woods. Also, there is no concern if saw teeth accidentally cut into the wedge.

Wood wedges are best made from hardwoods, and they may have a single or double taper. On a single-taper wedge, two sides meet to form a 90-degree angle. On a double-taper wedge, all three angles are acute. To make two single-taper wedges very quickly, fell a small hardwood with a diameter of around six inches and leave enough of the stump to work with. Make four straight vertical cuts around the stump to square it. Looking at the stump from above, make one diagonal cut (from one corner to another) that goes down about six inches. Then, make one horizontal cut through the stump at the bottom of the six-inch cut to release the two single-taper wedges.

The same technique can be done on a larger stump to make eight wedges. Once you've made the first diagonal cut, shift around the stump and make another diagonal cut connecting the opposite corners. Now that you have cut an X onto the top of your stump, make two more cuts from edge to edge like a plus sign. When you make the horizontal cut, you will release eight single-taper wedges. The experimentation is free, the wood is free, and you are likely to experience the sublime pleasure of making a very useful tool by hand (and saw). This is not chainsaw carving, yet you are still cutting an object from a large chunk of wood, guided by what you see in your mind's eye.

To make an Outdoorsman Ottoman, choose a tree with a trunk big enough to withstand multiple bore cuts and fell it by making notch cuts at about midchest height. Working at this level is better than stooping over to make the other cuts. Clear the fallen tree from your work area. The top of the stump where the notch cut was made will have some splinters from the hinge wood. Cut these off to make your stump level and to have a clean starting surface.

The Outdoorsman Ottoman is essentially a stool that is carved upside down. The legs are carved first, and then a horizontal cut at the end forms the top. The stool could be three-legged or four-legged, but I find the four-legged design the easiest to sketch out because you just need to draw a simple cross pattern on the surface of the stump. The cross must be as wide as your chainsaw bar. Using a stout ruler, mark off the cross with an ink marker. The four quarters outside of the cross will be the legs of the stool. All of the wood within the cross will be removed.

Next, take a rope and wrap it around the trunk, while keeping the rope level. Trace the ink marker around the edge of the rope. This line marks where the legs will meet the underside of the stool, and it marks the depth where you should stop cutting. You can make a short-legged stubby ottoman, a medium-legged coffee table, or even a high-legged bar stool. Use the rope technique again to trace a line for the final cut that creates the tabletop surface of the stool. The key to making this cut is keeping the bar level; otherwise the tabletop will be slightly slanted, and the imperfection will be obvious when you set something on it. One trick is to balance a torpedo level on the horizontal bar before making the cut to make sure the bar is level. Then, partway through the cut, use the torpedo level again (with the chain brake on) to ensure that the bar isn't deviating.

To make a Jim-Cut chair, you need to find a tree with two cojoined trunks that meet at a comfortable and safe working height. Make a peace sign with your index and middle fingers to resemble the two trunks of the tree. Imagine that your upright index finger will be the seat of the chair—this will be cut low near the base. The middle finger—which is at an angle—will form the backrest of the chair. Look for these Y-shaped trees where the Y is close to the ground. Typically, one of the trunks will have a lean and the other will be a little straighter. The trunk with the lean will form the back of the chair and the straight trunk will form the seat of the chair.

When you are felling the cojoined leaders, your felling cuts on the seat can be made low near the stump, but cuts on the trunk that forms the backrest must be made high. It's best to be prudent and make the cuts in such a way to leave more than enough wood to work with. Wood can always be sliced down to size later.

Once the trees are felled and a peace sign–shaped stump is left, you can make your Jim-Cut chair with just three cuts: one for the back, one for the seat, and one for the base. The cut for the chair backing is the most important cut. If the cut is too vertical, sitting in this straight-back chair will force you to adopt a ramrod posture and might be better suited for church or a schoolhouse. The aim is to capture angles similar to that of an Adirondack chair, which is the most perfect campfire furniture piece ever invented. The angles you cut will depend on the lean of at least one of the trunks; the farther the lean, the wider the angle you have to work with for the back of the chair.

Cut the back of the chair freehand and at an angle. To maintain a

Rather than cutting a Jim-Cut chair off at the base, you may just leave it as is and have a resting spot in the backyard or along a hiking trail.

straight cut, mark a line on the face of the flushed stump and a long, angled line that stretches from the top down to the base of the seat. A chalk line or a yardstick and a marker are helpful tools for this step.

It is important to make the angled back cut as level as possible all the way through the cut. Imagine sitting in a chair where one side of the chair backing is kinked forward and the other side is pulled backward. Upon sitting in such a chair, one shoulder would be hunched forward and the other would be slouched down and back. If your first try at the chair backing ends up like this, you can shave off a bit of the thick side. Before making that first cut, however, you must make sure you leave enough wood to work with in case the cut gets off-kilter.

I made my first Jim-Cut chair from an ash killed by emerald ash borers. Everything was going well until I made the cut for the seat and discovered some rot where the back of the chair met the seat. Upon making the final cut to form the base of the chair, I flipped the chair over and observed that the rotten area was even larger than I had expected and formed a football-sized hole filled with black ants. This was very disappointing, but not surprising. If the hole in the seat was any bigger, it would have looked like a primitive redneck toilet. The crotch of a cojoined tree is its weakest part, and debris often collects there to decompose. Also, insects regularly make homes in there. It is a common sight in the woods to see a cojoined tree's leaner crashed to the ground and the straighter trunk still left standing.

I sprayed out the ants with a garden hose, and I nailed some leftover treated-wood decking to the base of the chair. I cut the decking with a Hackzall, so it was flush with the contour of the base. That covered up the rotted hole, and the treated wood formed a layer of protection from the ground. I sprayed insect-resistant expanding-foam insulation into the hole. When it oozed and hardened at the top of the seat, I scraped it off, dug out some of the bulged-out foam, and patched and leveled off that area with wood putty. After applying a wood preservative stain to the chair, the wood putty patch was noticeable—but it was better than the gaping hole, and the project was salvaged. Jim-Cut chairs are heavy pieces of outdoor furniture that can be left outside for a long time before they deteriorate. Once one does, just toss it in the campfire and make another.

10

I Sing the Saw Electric!

The Battery-Powered Saw as a Stealth Machine

Do woodcutters dream of electric saws? I once did. I was an early adopter of battery-powered chainsaws. In 2015, I got an EGO brand saw with the idea that if I did not have electricity during a blackout, I could still power the chainsaw indefinitely via a small solar panel, which would keep me in the firewood game. I soon learned that it had two identities: it was often the first saw I grabbed because it was so handy for light work, but it was the very last saw I grabbed to do any serious cutting.

Battery-powered saws are extremely useful for cutting small wood. For many homeowners with leafy backyards, this type of saw is a great choice. For loggers, arborists, forestry professionals, and high-volume firewood cutters, the battery-powered saw is a useful addition to the toolbox but cannot replace the more powerful gas-powered saws. Currently, the battery-powered saw excels as a pruner, a limber, and for bucking small-diameter stove wood. But when it comes to speed and power, there is no comparison.

To its credit, however, it is a chainsaw of few parts. Bar oil is the only product that needs to be added. There is no on/off switch or pull start—you just grip the throttle interlock and another grip safety button, and then the chain zips to life. For people who do not understand combustible engines, this is the route to go. These saws are also extremely useful for tree-climbing arborists. I was once aloft in a big box elder and could not get a gas-powered saw started because the jerking motion had me bouncing around the branches. Eventually, the saw popped off, and I was able to make the pruning cuts. Many times, as a groundsman standing outside the drop zone, I have witnessed the same thing happen to other climbers. It is exhausting to mess with a recalcitrant chainsaw—especially while hanging by a rope thirty feet in the air. Usually, the stubborn saw is lowered to the ground and a more cooperative saw is roped up. But an electric chainsaw allows you to avoid this issue.

Weight-wise, battery saws are comparable to gas-powered saws, with the battery making up much of the weight. The battery, heavy as a brick, makes the saw very awkward, blocky, and end heavy. Gas-powered saws have superior ergonomics and are elegantly balanced. For quick backyard cutting tasks, the battery-powered saw will not pose a problem. But for those who work with saws all day, it is an issue of great importance. It is also a problem for those working in all weather conditions. Battery-powered saws are not recommended for use in the rain, whereas I've run gas-powered saws in steady, all-day rainfall and have even cut trees submerged in creeks.

Battery life can also be a problem. A single charge on a newish battery will get a medium-sized, blown-down softwood tree limbed and bucked up. Cutting into hardwoods, however, eats up more power and so does using dull chains. Lithium batteries and freezing temperatures do not mix, and much of my firewood cutting is done during the winter. I store my batteries—for all of my power tools—inside during the cold months. This can be a problem for mobile crews of arborists and loggers working year-round in the field and leaving their saws in work trucks overnight. For these workers, the best option for storing and recharging batteries would be for the foreman to round them up at the end of each day and charge them back at a workshop, home, or hotel, if working on the road.

Charging can take around a half hour to an hour, depending on the saw. And waiting, of course, slows down productivity. So, working in the field all day with a battery-powered saw requires either a fleet of spare batteries or a portable generator. If there is a power source nearby, any dead batteries can be charged on a rotation, allowing work to continue. But lithium batteries can be more expensive than the power tool itself. When my original EGO battery crapped out after a few years of use, I ordered a replacement online. It turned out to be a counterfeit and would hold only a minuscule charge. These scams are becoming more common every year,

Electric chainsaws are excellent for pruning and backyard tasks and make a solid choice for most homeowners. However, they have yet to surpass gas-powered saws for sheer power when cutting into big wood.

yet I have never purchased counterfeit fuel at a gas station. Likewise, unlike a jug of gasoline that gets emptied throughout the workday, the batteries do not become lighter for the hike out of the woods.

I also prefer some aspects of gas machines over electric ones. To determine how much fuel a gas-powered chainsaw has left, you can either look at a small window that shows the fuel level or unscrew the gas cap and take a peek. Gas saws also make a gurgling, dying sound before crapping out. To determine the remaining battery power of an electric saw, you must consult a meter that represents power in the form of illuminated bars. In my experience, these calculations are very vague until the saw suddenly quits.

T. J. Howell couldn't use electric saws in his juggling act because they didn't "bark" or have that loud revving noise to get the crowd worked up with anticipation. They were simply too quiet. But for the rest of us, this is the saw's greatest asset. Gas chainsaws are noisy as hell; electric chainsaws emit a whirring hum, like a swarm of enraged hornets. The noise takes on a deeper tone when the saw cuts into wood. Yet, there is no idling noise. As soon as you let off the trigger throttle, it is perfect silence.

I like to tackle the hardest work of the day in the morning, and sometimes that is chainsaw work near the house. This can be an issue for early risers when the rest of their family is sleeping. A gas-powered saw will let the entire neighborhood know what you're up to, but an electric saw is much stealthier. This can also be useful when you're trimming or felling trees along a property line with a nosy parker of a neighbor. Many folks can get worked up about their neighbor messing with border trees, even if it's a landowner's legal right to do so. A quiet electric chainsaw will get the job done with discretion and without a busybody interfering with the work at hand. I also use the electric saw when I want to make some quick cuts before deer-hunting season—usually it's a recent blow-down that obscures my line of sight from where I intend to hunt. With the battery-powered saw, I can slip into the woods and dismember the view-blocking tree with little disturbance to the wildlife.

California governor Gavin Newsom recently signed legislation banning the sale of gasoline-powered outdoor power equipment such as lawn mowers, leaf blowers, weed trimmers, chainsaws, and power washers in the state. The idea is that battery- or electric-powered equipment will replace the gas-powered machines, which will be better for the climate. Following suit, the Minnesota House introduced H.F. 1715, which bans gas-powered outdoor equipment in the state by 2025. A similar proposal was floated in Colorado.

In Jack London's short story "In a Far Country," he describes a couple of men whom he calls the "Incapables." They are characterized by pettiness, greed, and laziness, and they do not fully grasp the realities of surviving the Alaskan backcountry. I would argue that the politicians who proposed these initiatives in California and Minnesota do not fully grasp the realities of the difficult, labor-intensive jobs that involve gas-powered equipment. The workers who use these tools would be forced to perform hard work with underpowered equipment. Someday in the future, an electric saw may have the power of a gas-powered one. But for now, that's still just a dream.

11

Beautiful Slabs

The Saw as a Woodworking Device

This is a story about a man using a chainsaw to make his dream come true. It takes place on seven and a half acres of hilly land near Wittenberg, Wisconsin. Ryan Schlegel and his wife, Jen, bought the land in 2014 after looking at seventeen other properties. They drew a circle on a map and considered anything within a three-hour drive of their home in Stevens Point. "I was looking for woods," Ryan said.

The Schlegels and their kids spent years camping in many of Wisconsin's state parks. They loved the experience but did not love scrambling to reserve campsites that booked up in the blink of an eye. They wanted an outdoor getaway that was available whenever they desired it. The first thing they did on their new land was build a campfire pit. When I visited the property, I parked nearby on a horseshoe turnaround. I walked by the fire pit, some rough-cut benches, and a camper parked next to some trees. The Schlegel land included a hillside meadow surrounded by forest.

The land was once owned by the Tigerton Lumber Company. It is in a tension zone, an area once called the Gateway to the Pineries, where the great evergreen forests of the north blend with the southern broadleaf forests. In the 1800s, the Menominee and Ojibwe people living in the area were forced off the land by the US government. The government then sold plots of land to European immigrant settlers, many of whom deforested the land for agricultural purposes. A family with the surname Mavis bought 240 acres, and one member of the Mavis clan still lives down the road. Take 240 acres and divide it by four to get 60 acres. Divide that by four to get 15 acres. Half of that is how Schlegel came to own his 7.5 acres. It is the old story of large tracts of family land getting sold off piecemeal over the decades.

A few months before my visit to his country property, I was introduced to Schlegel at the Midwest Renewable Energy Fair in Custer, Wisconsin. It's the nation's longest-running event of its kind. Over the years, attendees have included redneck woodstove tinkerers, solar-power devotees, granola hippie types, biofuel moonshine distillers, windmill makers, and off-grid homesteaders. Schlegel and I sat in on a seminar about Urban Wood Network, an arboriculture group that takes big, beautiful trees cut down in cities and puts them in the hands of people who will turn the saved wood into something equally beautiful. Later, Schlegel and I started talking about chainsaws, and he invited me to visit his backwoods retreat.

Schlegel studied film at a liberal arts college in Santa Fe. He had an environmental bent—Edward Abbey is his favorite writer—so he decided to get into underwater photography. He moved to Florida, but during his advanced scuba training, he realized deep diving wasn't for him. He worked as a cameraman in Miami's commercial television market before moving back to Wisconsin in 1999 to be closer to friends and family. He worked in Milwaukee's film scene, and he did a lot of camera work for hospitals. "I was in so many hospitals, I stopped caring about what side somebody's hair was parted on in a commercial and started being more interested in what the nurses and doctors were doing," he said. He wanted a more portable trade, so he changed careers and entered the medical field when he was thirty-five years old. He is now a case manager at a hospital in Stevens Point.

On the day of my visit, the autumn colors were almost peaking, and Schlegel's woods were awash in orange, yellow, and red. Fallen maple leaves had already formed an ankle-deep carpet on the ground. Schlegel and I walked through the soft leafy slush. After all of the virgin timber was harvested by loggers, the land was used for grazing by the early settlers. Schlegel pointed out an old logging road gouged into a hillside ravine and a wall of piled rocks bordering one side of his land. He pieced together the history of the place and showed me the remains of an abandoned maple syrup operation. There was a large open tub for boiling down and evaporating the sap, along with assorted rusty metalwork. Schlegel taps his maples in late winter and harvests a couple of pints each year.

Since Ryan and Jen own fewer than ten acres, they were not eligible for Wisconsin's Managed Forest Law program, which offers reduced property taxes as an incentive for sustainable forest management on private woodlands. Instead, they hired a young couple who had recently graduated from forestry school. The couple walked through the Schlegels' land and marked trees to either keep or remove. But during this work, an accident happened. They felled a maple they deemed hazardous, and as it crashed, it took down a mature and healthy hickory tree that was one of the best ones on the property.

"I saw this hickory lying on the ground and I said to myself, 'I can't let this go to waste.' That was the beginning of my journey of getting into milling," Ryan explained. He estimated the trunk was around thirteen inches in diameter. Having heard about Alaskan chainsaw mills, he did some research and bought one. He didn't have the cash to buy a bigger chainsaw, so he talked a buddy into buying one with him and sharing the mill.

A chainsaw mill is a metal contraption that attaches to the bar of a chainsaw. Most can fit in the trunk of a car and are not too heavy to move about. The purpose of a mill is to guide the bar through the log at a predetermined depth, resulting in boards of consistent thicknesses. Large chainsaws are needed for this work—70 cc and above are ideal. Mills are made in different sizes, and there are only a handful of chainsaw mill manufacturers in the market. The leader in North America is the Alaskan Mill made by Granberg International.

There are just a few ways a person can take their own logs and turn them into useful dimensional lumber. Having the logs hauled to a sawmill is the easiest and most expensive option. You are simply paying someone else with highly specialized equipment to do the work for you. The next option is a portable sawmill that uses bandsaw blades or swing-blade circular saws. Calling some of these mills "portable" is a stretch. They are large machines and must be assembled and disassembled at the designated milling location, but some are permanently mounted on trailers, which enables them to be ported anywhere a truck can drive. The price of these mills ranges from $3,000 to over $100,000 depending on what industrial level you desire and how far you want to push your milling ambitions. Many owners of portable sawmills use them as a hobby business; they can be hired to mill logs on their customers' property. Milling logs is hard but immensely satisfying work. Also, by milling lumber for others, these owners are able to recoup the cost of the pricey machine and keep it maintained.

On the opposite end of the technological spectrum, another way to make round wood into straight-edge wood is to do it by hand with an axe or a crosscut saw. Before sawmills, the earliest method was hewing a log with a broadaxe. The hewer secures the log, then scores and chops it along a line marked in chalk. Once the plank has been cut from the log, the rough-hacked board is typically smoothed out with a drawknife or a hand plane.

Wood can also be milled by hand using a pit saw. Picture a log secured over a deep trench, supported by a couple of planks. One sawyer stands above, either on the log or on the planks, and another sawyer stands down in the trench. With each sawyer holding one end of a two-man crosscut saw, they push and pull the saw through the log. The pitman eats sawdust, and the sawyer above spends the day doing a backbreaking motion that resembles pumping air into a bicycle tire.

One method requires expensive and specialized modern machinery; the other requires tedious and exhausting manual labor. The middle-ground solution is the chainsaw mill. Chainsaw mills cost between $100 and $1,000, depending on their complexity and features.

Milling logs into lumber is another example of how the chainsaw provides agency and independence to the individual. By harvesting firewood with a chainsaw and using the wood to heat a home, one operates independently from the oil, gas, and electrical monopolies. When a storm-fallen tree blocks your roadway, you do not have to call the county and whine-and-wait like most people. You fire up the saw, cut up the tree blocking the road, and get on with life. When you mill your own lumber, you are no longer beholden to the bland and overpriced lumber offerings at big-box home improvement stores. You cut the size you want at the thickness you want from the tree you want. You operate outside the system of the building material conglomerates. For Ryan Schlegel, it was particularly important to mill lumber from trees on his land. Milling lumber from any other trees just wouldn't be the same. The connection to the land would be lost.

Most sawyers look at fallen trees differently after purchasing a chainsaw mill. Instead of eying them up for firewood, they consider the possibility of turning logs into live-edge slabs, boards, posts, and beams. PHOTO COURTESY OF JEFF VAN DE WALKER

We hiked farther down the hill to one of the nicest outhouses I've ever seen. It sat on a gravel pad edged with rocks, and behind it, a rain barrel collected water off the roof. It had windows, curtains, and shelves. "I made hardwood flooring for our outhouse," Schlegel said. "That's the first thing I did with my mill. We sawed into the hickory at about an inch and a quarter. When you're camping up here, [this outhouse is] the best place. You can sit in here and read a paper and enjoy life." After processing the downed hickory using the chainsaw mill he bought with his buddy, Schlegel brought the boards to a maker's space in Milwaukee and used a planer and joiner to finish the wood. He cut two grooves and added spline, a strip of wood to join the boards together. "The character of the wood came out fabulous," he said.

His long-term goal is to build a cabin on the land using wood harvested from the surrounding forest. Schlegel has been working his way up to it: first the outhouse and then the shed. On the hike over to his shed, we passed an apple tree. The ground was littered with recently fallen apples. Bees swarmed over the sweet windfall, and a ruffed grouse flushed from another nearby apple tree. After studying the size of tree rings on ash stumps near that apple tree, Schlegel estimated that grazing was halted on the land about forty years ago. His forest is mostly composed of maple, black cherry, hickory, basswood, and oak—and more red oak than white oak. Only a handful of pine trees are scattered on the land. Ryan plans to transform the meadow—which a deer hunter had once used as a food plot—into a hillside orchard and garden.

With so many maples on his property, Schlegel could end up harvesting some of the most prized and sought-after wood available in the region. Through milling, a sawyer can capture the beauty of spalted maple. Spalting is when fungus discolors wood by changing its pigmentation—essentially "bleaching" the wood through white rot. Competing fungi create strange lines in the wood as they mark their territory. Many people appreciate these chaotic streaks in the wood grain. This organic process of Mother Nature's artistry is from the right combination of moisture, oxygen, temperature, and time. The trick is to find and capture the wood before decay sets in too deeply and the wood becomes punky. The same goes for ambrosia maple, a beautiful wood with odd coloration formed by ambrosia beetles. Other beautiful types of maple include fiddleback maple, which has a unique rippled grain pattern; the rare quilted maple, whose grains resemble a wavy patchwork quilt; and birdseye maple, which has tiny knots speckled through the grain. The chainsaw is the best tool to capture the natural beauty of these rare specimens, and Schlegel believes he has good odds of coming across some of that highly coveted maple for his future cabin.

He designed his doorless shed so he could drive his Arctic Cat ATV in one end and out the other. However, the shed would not be doorless for long. Schlegel already had wood milled and earmarked for making sliding barn doors. A local Mennonite carpenter made the trusses. Schlegel and

Home to an ATV and hundreds of hanging garlic bulbs in the summer and fall, Ryan Schlegel's shed is basswood siding that he milled from a tree on his land.

his wife grow a huge amount of garlic each year, and the shed is the perfect place to hang the bulbs. He covered the structure with basswood siding he produced with his chainsaw mill. He did not apply stains or preservatives on the siding, and it will become weathered like old barn boards. It was a simple and sturdy shed—something that all back-to-the-landers need.

Not far from the shed, we walked to the sunniest spot on Schlegel's land. He was using an old trailer parked there as an off-the-ground platform to season the milled boards that were sticker stacked for air drying. He snatched a board off the trailer. "This one is really gorgeous. I can't believe how pretty that turned out. That is straight as can be," he said, flipping it upright and eyeing it lengthwise. "This is all oak that I milled from a tree that fell last year. This stuff is two inches, which is the sweet spot, I think, after all the milling we've done. The two-inch cut is probably the best thickness for stability. If you go any thinner, the wood wants to twist. But if you keep it two inches and dry it that way, and you send it through the planer a dozen times, you end up with some really nice flat wood that you can aggressively sand, and you don't end up with a thin sandwich board. Over the years, we found two inches is about the right cut. It gives you enough wood to play with, and it keeps the wood stable while its drying."

Schlegel explained that he keeps a garage in Wittenberg, where he stashes milled wood that he will eventually use for his cabin. Past the meadow, we walked into a boulder garden in a wooded swale. This part of the forest had a different feel to it—it was peaceful, pretty, and eerie at the same time. Smooth rocks left behind from the last ice age dotted the forest—a common feature in this region where the last glacier stopped moving and started melting.

I noticed that the line of sight in this area was disrupted by windthrow trees and widow-makers tangling up the halcyon woodscape. "There was a tornado outside of Wausau. This is the damage from this storm a year and a half ago," Schlegel explained. Farther up the hill, near the rock-pile wall, he pointed out a fallen basswood tree taken down by a tornado in Amherst two years earlier.

"The idea was to take what's on the ground," he said of his early milling strategy. "Don't let it go to waste. Not that it would, because the forest would do a good job with it. After that first hickory, basswood was the next tree I milled. The hickory was a bitch to get through. With the basswood, it went through it like butter, and we didn't have to stop and sharpen the chain as much. I made a couple benches. I put those benches around the campfire pit. Years go by and there is no insect interest at all. They are still solid as the day I cut them up."

Schlegel has never moved or skidded any logs to a milling site. He has always milled the timber exactly where the trees fell. That is the beauty of a chainsaw mill. It can easily be taken to remote areas, whether a roadless area accessible only by bush plane, pack mules, or boat or a nearby but inaccessible area such as a wetland or a steep ravine. With a chainsaw mill, the wood can still be captured and saved. In addition, if a log falls in a problematic spot, Schlegel can adjust it into place with the use of a chainsaw winch.

The Lewis Winch, a portable winch that uses the chainsaw's powerhead for cranking energy, is another example of a chainsaw adaptation that empowers the individual sawyer. As long as a sawyer is using a saw with five to seven horsepower, this winch makes it possible for a one-person operation to skid heavy logs into place in remote areas without the need for heavy machinery.

The Lewis Winch was invented in the 1970s in New Westminster, British Columbia. The newest model is made of an aluminum alloy and is strung with aircraft cable. With an adapter kit, it can fit on almost any saw on the market. Once you attach the saw powerhead (and remove the bar and chain), you simply wrap a chain or tow strap around a tree or some other solid anchor and attach that strap to the winch with a clevis. Next, free-spool out the cable to the log that is cinched up with a timber hitch or another chain, hook it up, and hike back to the saw-winch. Finally, once you start the saw and goose the throttle, the winch reels the log right to you. The Lewis Winch company also makes a multidrill attachment for chainsaws. With that device, a lone operator can drill holes into the earth for posts or tree planting. It can also be used as an ice-fishing auger.

At Schlegel's current milling spot, I spied his Echo CS731 OP with a twenty-seven-inch bar mounted to his thirty-five-inch Alaskan chainsaw mill. I recognized the skip-tooth chain, which is used for milling. On these chains, every other tooth is missing, and because there are fewer teeth and longer gaps along the chain, the skip-tooth chain has better debris and wood-chip clearance than standard chains. Semi-skip chains have a mix of skip and standard teeth. Full skip chains, which have the biggest spaces between the teeth, are used for hardwoods and big slabs. On small wood, that type of chain is grabby and unsteady. Because skip-tooth chains have fewer cutting teeth and therefore less drag, they cut faster than normal chains. The spacing between the teeth clears wood chips faster, and the decreased vibration makes for a smoother cut.

"A lot of people wonder how you make the first cut," Schlegel stated. "Essentially, you need a flat surface for the mill to run on. Granberg [International] created this jig to make the first cut, but at first, we bought two twelve-foot cedar posts. We attached them with little jigs that they sell, and we used these posts to create our first-cut rail system. Other guys online use an aluminum ladder. This fancy thing from Granberg, or a homemade job with some straight lumber, or an aluminum ladder—it doesn't matter what it is, you need one of those three things in order to start that first cut. The saw is set much deeper for the first cut. After the first cut, that's when you start setting your depth for whatever you want. So, your saw is running, you're . . . very careful, and then you get into a good place, get a nice start, and boom—you're going. Once you get into the wood, it's a much safer operation. You want to work downhill. The gravity will work in your advantage a little bit."

He showed me a photo from his phone of some three-inch slabs he cut with the idea of using them for countertops. "Pretty much everything I cut for future interior work will be two inches thick to compensate for all of that planing and sanding," he said. Then, he looked around the woods. "We don't have a lot of pine. It will be mostly maple and oak. So it's mostly going to be finish wood. It's going to be interior—cabinet doors, countertops, shelves. It's been super fun. It's a no-brainer. It just takes time."

12

An Elegy for the Ash

The Saw as a Tool for Cutting Our Losses

I woke up early and had a breakfast of two over-easy eggs; venison tenderloin wrapped in bacon; a wedge of buttered cornbread; a sautéed side medley of onion, garlic, and jalapeños stirred into a plate of corned-beef hash; one pot of coffee; and a slice of warmed apple pie topped with a slab of melted cheddar. This is the first law of thermodynamics. Energy can neither be created nor destroyed, just changed from one form to another. I would burn thousands of calories today cutting wood, and in turn, I would produce millions of British thermal units that would warm my home in the form of firewood the following winter.

The previous night, I'd sharpened Katana and performed my regular saw maintenance ritual. I brought in my saw chaps to warm overnight by the woodstove. It was a cold January morning, but the sun would come out and warm the day. I loaded the saw, chaps, fuel, oil, gas, tool pouch, and a thermos of warmed water into the sled and hiked up the hill. Some days, the chainsaw is a tool for taking out the dead. The sawyer is the undertaker. When the Black Death descended upon medieval Europe, the dead were everywhere and piled-up bodies were hauled away in carts. It is now the same in American forests. Disease and pestilence are ravaging our woodlands.

The riverine zone of the Waupaca River is a horror show of dead ash trees. They stand like skeletons in the forest. Their bare upper branches reach past the green canopy of healthy trees like the bony fingers of a decomposing zombie emerging from a grave. Some of them *are* zombies.

They are essentially dead, yet there is still a bit of life in them. Green-twig suckers protrude off their main trunks—a desperate last photosynthetic chance to capture life-giving sunlight after the main canopies lost their leaves long ago. An ash stump of a tree that I felled a year ago still sends up suckers.

The emerald ash borer is an invasive beetle native to Asia that is killing ash trees across North America. There is very little that can be done to stop it, and I believe we are witnessing an extinction-level event for the green, black, white, and blue ash trees. The beetles kill the green and black first. After those are eliminated, they attack the white ash. The blue ash tends to be more unpalatable to the borers, and therefore slightly more resistant, but eventually they die, too.

The larvae chew through the inner phloem, cambium, and outer xylem of the trees, which then die slow deaths over the course of a year or two. It takes longer for bigger trees. A tree will lose its leaves and develop flecking on its trunk where woodpeckers hack at the outer bark hunting for the larvae. This leaves blond patches scattered up and down the tree. Once you see that, you know it is terminal. If you knock the bark off the dead trees, you can see the larval tracks. They leave behind serpentine trails that curve and twist. In a way, the patterns are weirdly ornamental and attractive, but once you know what they indicate, they are mostly depressing. I estimate that for the next few years, ash will make up the majority of my annual firewood inventory. I believe that someday an ash baseball bat might be a rare treasure. There is a very good chance that a child born today will never see an ash tree in the forest as they grow up.

The dead ash need to be taken down. When I worked as a line-clearance tree trimmer, my company had a policy that we could not climb ash trees even if there were green leaves on them. They were so likely to be unstable that it was not worth the risk. Just about the only good thing to come out of this catastrophe is that ash trees make excellent firewood. The wood splits easily (although it can be stringy sometimes), and it burns well. In *The Firewood Poem*, written in 1930, Lady Celia Congreve lists all of the types of wood that can be burned for warmth, mentioning ash three times—once at the end of each stanza. She writes, "But ash new or ash old, / Is fit for a queen with crown of gold," and then, "But ash green or ash brown / Is fit for a queen with golden crown," and finally, "But ash

The curlicue striations on this ash log are the tracks of emerald ash borer larvae as they chew their way through the inner bark.

wet or ash dry / a king shall warm his slippers by," emphasizing that ash is firewood fit for royalty.

Once I reached the top of the hill, I tackled the most difficult tree first. The ash was what is called a schoolmarm, as it forked into two leads just above my head. After circling the tree, I realized it leaned equally both ways. These trees can be unpredictable. A similarly shaped tree is called a devil's fork, a type of evergreen with multiple tops, like a bonsai tree with branches going in every direction. It's tricky to judge the head lean on trees like this. Both leads of the ash had a slight downhill lean, so that was the direction I would send it to the ground. There were plenty of openings in this felling zone, and it did not matter where it landed except for one spot where a red oak stood.

This oak was downhill from the schoolmarm and had witness wires grown into its trunk; two rusty strands of barbed wire protruded from the scar. Some old trees have unnatural gouges on the trunk at chest height, which means they might be what are called bearing or witness trees. These trees were marked by early American surveyors to plot the lines of townships. Confirming my theory, when my neighbor had the property line surveyed, the survey markers lined up exactly with this barbed-wire oak. Alternately, the cuts could have been made by a timber cruiser or a private surveyor marking trees for harvest or determining private property lines. The marks were likely made by a timber scribe, a tool that looks like a sheath knife with the tip of the blade bent into a semicircle. The curved end is used to gouge deep cuts into tree bark. Spray paint has now replaced this forgotten tool.

I pressed my chainsaw bar against the trunk of the ash, aimed my felling sights, looked up at the canopy above, and looked back at the sights. The ash would fall clear of the red oak. I absolutely did not want the slightest harm to come to the oak, even though it was safe at this time of the year for an oak tree to take on some damage. A handful of old, majestic oak trees have died on my property from oak wilt, a disease caused by the fungus *Bretziella fagacearum*, which is transferred by beetles that carry its spores. To prevent the spread, I knew I should not cut or prune my oaks between early April and mid-July. If cutting was necessary, I would apply pruning paint to the cut so the beetles would not be attracted to the sap.

Death by oak wilt comes quickly, and a big oak can be leafless within a month. Earlier this year, I drove by a stand of red oaks and noticed that the biggest one had no leaves. I shut off my four-wheeler and stared at it, flabbergasted. Earlier in the spring, it had been a healthy mature tree. Now, it was another dead tree to take down. I did not ask for this. Oak trees are more valuable to me alive than dead as they provide acorns for the wildlife in the woods. I have developed a felling order of trees to be removed in my woodlot. This tree-removal triage changes from year to year according to disease and destruction. I keep a running list in a small Rite-in-the-Rain notebook that, fittingly, holds all-weather paper invented in 1916 for loggers in the wet Pacific Northwest. I have very few elms on my land, and the biggest one died three years ago from Dutch elm disease.

The tell-tale signs of a dying ash are pale blotches that appear on the trunk. Called "flecking" or "blonding," the discoloration is from woodpeckers hacking away at the bark to get at the emerald ash borer larvae.

The list of dying trees continues to grow, and the ash is currently at the top of the leaderboard.

I swamped out the area around the base of the ash, nipping away small vines and springy buckthorn seedlings and pulling and tossing dead branches that were tripping hazards along my escape route. I cut a conventional notch into the ash and eased into the back cut until there was just a thin strip of hinge wood left. Then, I let off the throttle. There was a barely perceptible sway of the trunk. In this moment, the tree now belonged to gravity. I backpedaled uphill on my escape route as I watched the ash hit the ground. The dry, dead upper branches shattered upon impact. The red oak remained untouched. I limbed the trunk, then continued limbing up one fork of the schoolmarm and then the other.

I took a moment to study the bare log. I wanted to prevent the most aggravating predicament in chainsaw work: a pinched bar. It happens to novice chainsaw operators, but it also happens to experienced sawyers who are working too fast and not being mindful. When a bar gets pinched in a log, sometimes a sawyer can rock the wood back and forth and pull the bar out, but you do not want to damage the bar. In any circumstance, a stuck bar is frustrating. Your productive work session suddenly comes to an end. All you can do is turn off the engine, take a break, try not to swear too much, and develop an extraction strategy. One method is to get another saw and make a cut that will relieve the vise-like pressure. Another way is to hammer wedges into the kerf and open it up enough for the bar to be removed. If a friend is around, you can ask them to sit or stand on the log, or lift or push it at the leverage point, in order to open up the kerf.

A pinched bar is the result of tension, compression, and gravity. Top bind describes when a bar is pinched during a top cut. For instance, if you were bucking a log positioned bridgelike over a gulley, the compression would be on top and the tension would be on the bottom. If you cut from the top, the log bridge will sag and pinch the bar. With bottom bind, the situation is reversed. In the case of a log with one end sticking up off the ground, an undercut will result in the bar being pinched from the freed part of log that wants to drop. It's best to cut it at the hinge point where the log sticks up from the ground.

A no-bind log—like a log lying on flat ground—is easy to cut, as it doesn't have any competing pressures. End bind is exemplified by a log on a steep hill; if the sawyer cuts on the upper or middle part of the log, the weight of the uphill log presses into the kerf and pinches the bar. Better to start cutting on the downhill end and let the cut chunks roll away. Side bind is when a tree is tensioned on one side and compressed on the other, such as when a felled tree settles and wedges itself between two trees on either side. In this case, the pressure on the log comes from either the left or the right. The best course of action is to cut the compression wood first and to watch the kerf while cutting. If it starts closing up, pull the bar out of the kerf before it gets pinched. Then, continue cutting the compression wood at various intervals (firewood size or whatever length desired) until the compression wood no longer pinches the bar and the log can be cut through.

I cut the branches of my felled ash into firewood-sized chunks. Then, starting at the top, I bucked my way back toward the trunk until just one firewood-sized round stuck out from the stump, held by hinge-wood

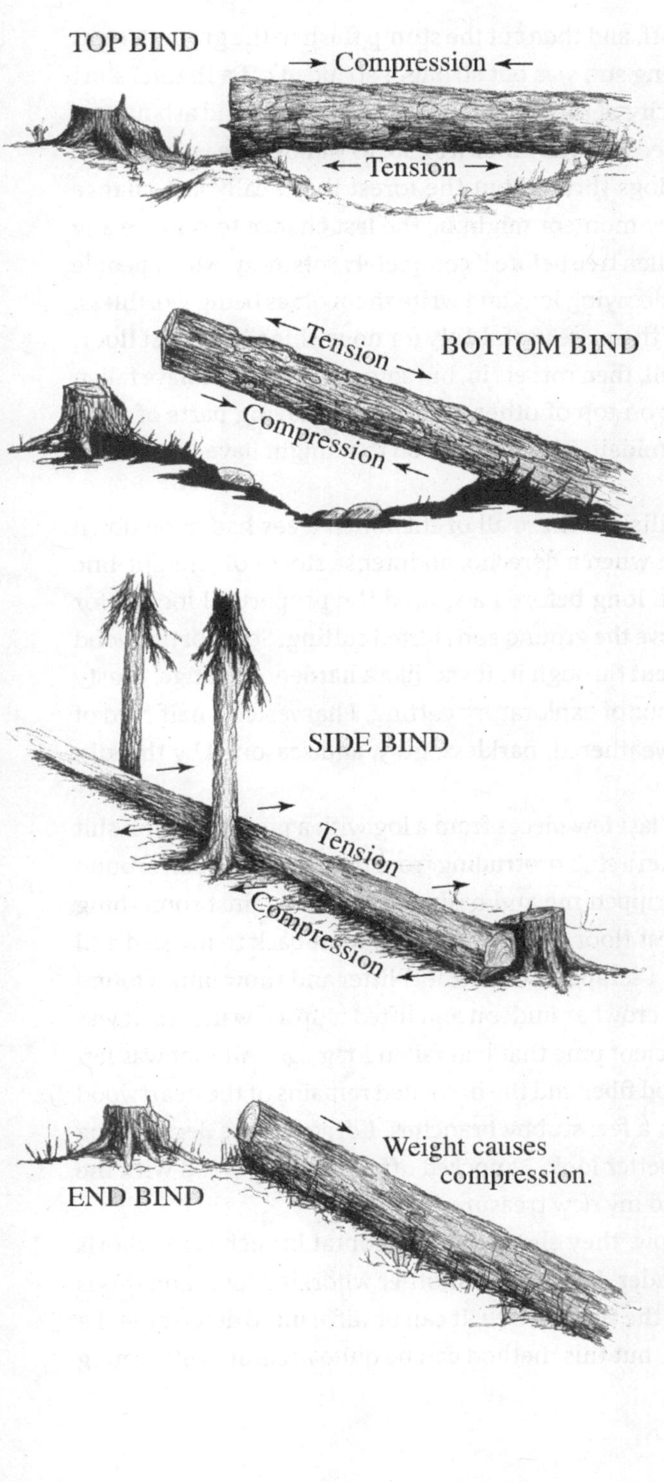

These are the types of logs that can "bind" or pinch a chainsaw bar. Gravity, log weight, and pressure factor into every situation, and a sawyer should study each log before making the first bucking cut. *USDA FOREST SERVICE CHAIN SAW AND CROSSCUT SAW TRAINING COURSE*

fibers. I sliced these off, and then cut the stump flush to the ground. I was sweating. The morning sun was out strong. I stripped off a flannel shirt and enjoyed the apricity of the day. I felled another half-dead ash nearby and limbed and bucked it. Then I took a break to wander the woods.

There are fallen logs throughout the forest that I call "last-chance wood," as in, this very moment might be the last chance to capture any firewood from this fallen tree before it completely rots away. Many people would walk by these decaying logs and write them off as being worthless. And it's true, some of them are useful only for nourishing the forest floor. If wood is touching soil, then rot sets in, but some of these trees have fallen over small gullies or on top of other fallen trees, leaving parts of their trunks suspended in midair. This is the wood that might have some value left in it.

I hiked along a hillside where all of the fallen trees had gone down in the same direction when a derecho, an intense storm of straight-line winds, came through long before I acquired the property. I looked for trunks suspended above the ground and started cutting. Some of the wood crumbled as the saw cut through it. It was like a hardened sponge, crusty and light. After an hour of exploratory cutting, I harvested a half cord of hardwood that was weathered, barkless, gray, and seasoned by the kiln of time.

While cutting the last few pieces from a log with a pile of raccoon shit on one end, I tripped on a stub protruding from the ground. I spun around to kick at whatever tripped me and bashed my boot against something hard stuck in the forest floor. It didn't move. I went back to my sled and grabbed the crowbar. I scraped away the leaf litter and snow until I found the log. I jammed the crowbar under it and lifted it up a few inches. It was the skeleton of an ancient pine that had fallen long ago. All that was left was some spongy wood fiber and the hardened remains of the heartwood and pine whorls with a few stubby branches. I dragged the decayed log onto a boulder for a better look. I knocked off the sponge wood with the crowbar and inspected my new treasure.

As evergreens grow, they shoot out horizontal branches in whorls around the central leader. Every year, another whorl is added, and this is one way to figure out the age of a tree. It can be difficult to determine the exact age of old trees, but this method can be quite accurate with young

Pine knots: the gold of the forest. Every home, hearth, and cabin should have pine knots and fatwood for fire-starting material.

ones. In between the whorls is a space of bare trunk. During times of good growth, this space will be long, and during dry years it will be short.

I started my saw and cut the log by separating the whorled branch junctures from the branchless bare spaces of the trunk. It was quick work and, once finished, I scraped the sponge wood away from the whorls to reveal a knotted, gnarled petrified pine chunk. It was very heavy in my hand, and it held the last of the tree's pine resin in its core. I call these pieces campfire candy.

One of my favorite short stories is "Nothing to Do for Three Weeks" by Gordon MacQuarrie, who became the first full-time outdoor writer in the US when he was hired by the *Milwaukee Journal* in 1936. The narrator in the story heads up to a North Woods cabin and preps the place before the arrival of his friends. He goes out to gather firewood and comes across the remains of rotted pine logs left behind from the lumber boom. All that is left are knotted whorls. "You cannot shoot a pine knot, or eat it, but it is a lovely thing and makes a fire that will burn the bottom out of a stove if you are not careful. Burning pine knots smell as fine as the South's pungent lightwood," MacQuarrie stated.

"Until you have your woodshed awash with pine knots, you have not ever been really rich," he wrote. The narrator gathers seven boatloads of pine knots by making two-mile round trips to the old lumbering spot. "In even the very best pine-knot country, such as this was, that is a tremendous haul for one day; in fact, I felt vulgarly rich." MacQuarrie ends the story with the narrator referring to himself as a "pine-knot millionaire."

I will never be a pine-knot millionaire, as my land does not have that many pines, but I have always sought to increase my pine-knot net worth. That day, I gathered the knots and set them in a pick-up-later pile near the trail. George Herter, the early sporting goods magnate and eccentric author, recommends doing the same in his *Professional Guide's Manual*, first published in 1960. "Look for a rotted out stump of a long dead pine tree," he wrote. "Kick the stump apart with your foot or knock it apart with your axe. You will find some pitch lumps toward the center of the stump. These pitch lumps will start a fire even of the wettest kind of wood." To me, the few pine knots I'd found were the small-scale woodland equivalent of discovering a seam of coal in an exposed rock or petroleum seep gurgling from the ground—a kind of hidden stored energy, there for the taking for those willing to put in the work to extract it.

I grabbed a round that was cut from between the whorls. It was free of knots and split easily, yet the wood, like the pine knots, had a hardened, petrified, and resinous feel. I split it open with a hatchet. The wood inside was a rusty orange, much like the highly prized pumpkin pine, which results when boards milled from old-growth eastern white pine develop a glowing orange patina after spending decades out of direct sunlight, like in the attic of an old farmhouse. In a way, this made sense as the carcass of the old pine had been resting under leaf litter for decades. The wood smelled of strong turpentine. This is what MacQuarrie referred to as the "pungent lightwood" of the South. Called fatwood by some, it is most commonly associated with the longleaf pine that grows in the southern part of the US, but fatwood can appear in any species of pine.

Not all of the old, petrified pine I found was fatwood. Some of it, when dried, would simply be normal kindling. I split other rounds, chopped off some kindling splinters, and stuffed them into my pocket to test later for flammability. At a hardware store, I once saw some expensive fatwood sticks in rustic packaging. I thought it was a scam. The sticks lacked the resinous luster of the fatwood that I harvested, and they'd been cut into perfect and characterless rectangular sticks. They appeared to be generic pine, which makes great kindling on its own, so the users would not have a clue if they were burning fatwood or not.

I was delighted by the fact that this gold of the forest had come from a tree almost obliterated by decay. Unlike MacQuarrie, I do not burn pine knots in my woodstove. Although they burn hot, they also release a black, sooty smoke, so I save them for campfires. There is a myth cautioning people against burning pine in a woodstove because it leads to creosote buildup that could result in a chimney fire. However, pine is fine to burn as long as it is well seasoned and has a moisture content of less than 15 percent.

Finding this long-dead pine made me happy, but there were other dead pines in the forest that left me soured and concerned. All of them were red pines; two were large, towering giants. I suspected heterobasidion root disease: a killer fungus that makes its way into the root system of freshly cut stumps and then transfers into the root system of other trees when their roots make contact. Yet it could also be red pine pocket decline or diplodia, two fungal killers that are similar to heterobasidion. There were not many pine trees in my woodlot, but I wanted to save what I had. However, to keep this death and dying in perspective, a few dead pine trees did not compare to the total annihilation of the ash trees.

I was most concerned about one tree in particular. I called her the Queen of the Forest, a large white pine with a robust canopy. She was a wolf tree that towered above a younger stand of poplar, birch, and black cherry. A wolf tree is a stand-alone tree with an abundance of sunlight and no competition. They're able to spread their branches wide and grow tall. Early foresters thought that these trees, like wolves, consumed too much of the forest resources (sunlight, nutrients, water) and ought to be culled. Sometimes the early settlers would leave some nut-bearing wolf trees in their pastures for shade, shelter, and mast for their livestock.

Between 1937 and 1941, the US Department of Agriculture and US Geological Survey took more than thirty-eight thousand aerial photos of Wisconsin. These images are now accessible online, and so I located my property. Other than some foliage by the river, the land was completely barren of trees. Too hilly for the plow, its inhabitants must have used it for pasture land, which explains how the Queen of the Forest was born. If she was alive at that time, she was just a speck of a sapling.

About one hundred yards uphill from the queen was a pile of birch rounds that needed to be hauled to the woodpile. The paper birch I cut a week earlier was one of three cojoined trunks that merged at ground level. Two of the trunks were healthy. The third, the largest, was dead with no explanation. Already, on its upper branches, decay had begun to set in, and fungus was growing on its bark. I did not know what caused this one to die—perhaps the bronze birch borer, another tree-killing beetle—but I did not want the other two trunks to suffer the same fate, as I prize the paper birch in my woodlot. One of the trunks had a catface, a kind of scar, caused by a windthrown silver maple that gouged its trunk. Now that impact spot was gnarled, deformed tissue. Catface is considered an imperfection among loggers. Down in the southern pine forests, one kind of catface on a tree can be the diagonal V-shaped slashes that direct and collect pine resin for turpentine production.

One of the reasons I love birch trees so much has to do with their relationship with mushrooms. I've only once found chaga (*Inonotus obliquus*); it was growing on a birch tree near a vernal pool that temporarily appeared during the spring snowmelt. The pool was filled with spring peepers, salamanders, and wood frogs. The chaga was at shoulder height, and I was able to cut off a crusty chunk with a Fanno handsaw. Chaga is highly valued for its medicinal uses. All endemic cultures that have existed where birch trees grow identify the fungus as a folk remedy—from Russians to Scandinavians to certain northern-dwelling Asians and American Indians. Chaga is rare, but the dead birch I felled on my land had some very common horse-hoof fungus on it. The inner tissue of this fungus (*Fomes fomentarius*) can be turned into a mushroom leather called amadou. I once made some by soaking the trauma layer in washing soda and pounding it flat with a mallet. It is flammable, and in ancient times it was highly valued as a fire starter. Some amadou was even found on the remains of Ötzi the Iceman, Europe's oldest known mummy. It is also highly absorbent, so I saved a patch for drying out waterlogged dry flies.

Having cut up some ash and gathered a small pile of last-chance wood, pine knots, and fatwood, I considered my morning productive. On the neighboring property, I could see jackstrawed poplar trees that had been blown down in a windstorm, leaving a crisscrossed tangle of widow-makers and unholy snags. With the new openings in the canopy, more invasive brush had moved in and made it a woody, debris-filled jungle—a kind of hell for people but a cozy and secure bedding area for whitetail deer. This unchecked spread of invasive brush, along with the death of the native trees, depressed me. In my conversation with Professor Demchik, we'd traded stories about how to handle buckthorn and other invasive species. I told him about my struggles. "That's going to be an incredible challenge over there," he responded. "The soil is a little bit better [than average] and that 'little bit better' means it's going to be even more favorable for some of these invasive species."

I fired up the saw and walked around the felling area in a small circle that spiraled into larger and larger circles, slashing down all autumn olives, buckthorn, and honeysuckle in my path. After fifteen minutes of nonstop cutting, I was still only thirty yards away from my work site. My goal was to have this spot cleared by spring so I could plant this patch of hillside with tree seedlings. In early winter, the county conservation department sends out its annual tree order form. Each year's catalog is a little different, with small price increases or new species of seedlings in the offering. The county sells seedlings in lots of twenty-five and a hundred. I am always excited to get this flyer, just as passionate gardeners look forward to receiving their seed catalogs in the dead of winter. I knew that having the site prepped for the seedlings would give the trees a head start.

These are the plagues ravaging the trees in my forest, but similar battles are happening across North America. Out east, beech trees are getting killed by beech bark disease, a malady that results from an insect opening up the bark and making a pathway for a fungus. The Eastern hemlock is dying from hemlock woolly adelgid, another destructive insect. Out west, pine bark beetles are laying waste to conifers and spreading rapidly. After the bark beetles carry in the fungus that kills the pine trees, the wood grain is left with weird bluish-gray streaks. Enterprising sawyers are working with local sawmills to salvage this eye-catching wood and get it into the hands of artisans and carpenters. However, the work is never-ending. And in this effort, chainsaws are needed more than ever. The world is full of dying trees, and their requiem is the sound of a chainsaw.

13

Slices for Science

The Saw as a Harvester of History

My ideal winter day ends with two hours of chainsaw work before sunset. I try to plan my workday so that those two hours are unmolested by obligations, and about half the time I succeed. Two hours of chainsaw work every other day adds up to a lot of firewood. The strenuous exercise and the ice-cold weather is a nice contrast to an evening that will be spent lounging and reading by the woodstove.

During one of these sunset sessions, I processed another dead ash. Once I piled the firewood by the trail, just the stump was left. I cut off the splintered top with its hinge wood and then an inch-thick cookie. I shut off the saw and sat down in the snow with the stump as a backrest. Then, I put the disk in my lap and began the autopsy. I already knew the cause of death—the emerald ash borer. I wanted to know the age of the victim. I counted the rings, but there were many spots where they blurred together from the rough cut of the chainsaw. Studying it after sanding it smooth would be far better.

By my estimation, the tree and I were about the same age, give or take a few years. It made me sad to think of the tree dying a slow death as the insects girdled its tissues. It made me sad that I had to cut it down in the first place, but there I was, salvaging what I could for firewood. I would plant a sapling in its place in the coming spring.

Dendrochronology is the scientific method of dating tree rings. The father of dendrochronology was Andrew Ellicott Douglass, an astronomer interested in how sunspots might affect Earth's climate. In 1904, he sawed twenty-five rounds off some ponderosa pine logs at a lumberyard in Flagstaff, Arizona. With those rounds, Douglass was able to make connections between solar activity, the Earth's climate, and tree growth. He pushed ahead with this research concept and collected samples of trees from California and the American Southwest. At the University of Arizona, he became the first professor to teach a course on dendrochronology and founded the Laboratory of Tree-Ring Research, which is now the world's largest dendrochronology library.

"Many people are comfortable with the concept behind dendrochronology, having looked at the top of a tree stump as kids and counted the rings," wrote Valerie Trouet in her 2022 book, *Tree Story: The History of the World Written in Rings*. "It is a tangible field of science: you can touch the wood with your own hands, you can see the rings with your naked eye. It does not involve obscure nanoparticles or far-flung galaxies."

Today, the main tool used by dendrochronologists is the increment borer. It's a T-bar-shaped hand auger that can be drilled into a trunk to remove a cylindrical core striped with growth rings. The tool is used on living trees when there is no reason for them to be felled. Crosscut sections of trees are often featured in nature centers and classrooms, which got me wondering about where the chainsaw fits into the toolkit of modern dendrochronologists.

I got in touch with Professor Evan Larson of the Tree-Ring, Earth, and Environmental Sciences Laboratory (TREES Lab) at UW–Platteville. The lab was founded in 2012, and its students and staff work on projects involving dendrochronology, geospatial analysis, and sediment-based quaternary paleoecology. "We've got a couple chainsaws," Larson told me. "We've done a good bit of work with them. We're working to translate the patterns of tree growth rings into a more complete understanding of how the world around us is changing and how it changed in the past so we've got a better idea of how it will change in the future."

Larson grew up using chainsaws in central Minnesota in a family home heated with firewood. Later, as an undergrad studying environmental science at Willamette University in Oregon, he was singled out as the perfect candidate for some forestry fieldwork. The reason? He knew how to run a chainsaw. He later earned his master's degree at the University of Tennessee–Knoxville and his PhD from the University of Minnesota. In his job at the TREES Lab, Larson mainly uses chainsaws to cut cross sections of logs to examine tree rings. Larson has also cut with a handsaw underwater to collect samples on sunken logs and pilings. He once pitched a proposal that required an underwater chainsaw, but the project didn't get funded. "The area[s] of dendrochronology where saws become important are the types of work where a core will not help you answer the questions that need to be answered," he said.

Chainsaws are most often used by Larson and his team when they're investigating fire history. Wildfires were once very common across North America. Many of the intense wildfires we see today completely destroy trees, but it's possible for trees to be scarred by fires and continue to grow. Some may resprout, and serotinous trees will even reseed—their dormant pine cones can open up after exposure to fire. Larson explained, "The forests that we see and were eventually logged—at first with crosscuts and later by chainsaws—were almost universally shaped in one way or another by fire. We [modern scientists] didn't see it because we didn't have the lived history and didn't understand the long-term role and the slow expression of these forest ecosystems."

He pointed out that all of the huge white pines logged in the Great Lakes region were able to grow large because fires wiped out their competition. The ponderosa pines in the west were born from and maintained by fire. Both wildfires and fires intentionally set by Native people helped keep the underbrush low, allowed serotinous pines to reseed, and killed smaller competing trees, which allowed the bigger ones to become giants. Native communities across the continent have used fire as the most effective tool to manage their surrounding landscape. The subsequent pioneer stories of settlers being able to drive their horses and carts through wide-open stands of huge oaks resulted from these fire-maintained landscapes.

Pine trees have a thick bark that protects them from low-severity fires. While such fires do not affect older trees, the younger trees are exposed to a swirling vortex of heat that cooks the cambium through the bark. As a result, the cells in the cambium stop dividing. As long as a portion of the tree survives the fire, the tree heals itself by sending resin to the injured area of the trunk. The resin does not include water, so wood decay fungi cannot grow. Eventually, this wound area will seal up. In the instances when it does not, the bark in that spot does not return, making the burn

A student from the University of Wisconsin–Platteville's Department of Environmental Sciences and Society uses a chainsaw to remove a partial cross section from a dead fire-scarred old-growth red pine (*Pinus resinosa*) at Mirror Lake State Park. An initial sample was removed at a higher spot on the trunk and appears as a dark horizontal slot. PHOTO BY JAMES P. RISER II

wound more susceptible to the next fire. When another fire occurs and cooks the cambium at the fire scar, the healing process resumes again. This process is repeated for years. Eventually, the trees become deformed, with the healthy side of their trunks looking round and normal and the burned sides caving in with fire-scar tissue.

It is difficult for an increment borer to cut through multiple fire scars, and the most important information in a sample is at the very end of the scar where dead cells meet surviving cells. When Larson and his team study the growth rings of these trees, the fire scars will precisely show when those cells stopped growing. At Mirror Lake State Park, Larson discovered a red pine that has fifteen fire scars on it, which means fifteen fires passed through the area during the tree's lifetime. "You can actually identify the season of the burn," he stressed. "It's 100 percent accurate. I'm not saying it was 1804 or 1805. I'm saying it was in August of 1804 that a fire burned by here."

Since it is extremely difficult for an increment borer to capture this valuable information, a chainsaw technique was developed to cut out what dendrochronologists call a partial cross section or a partial wedge, which resembles a very thick slice of pizza. "If you do it right," said Larson, "you get the pith and the whole fire record. You get the age of the tree, the date [the first ring formed], and every time a fire scarred that tree."

Stephen F. Arno and Kathy M. Sneck were the first to document the method in a 1977 report for the US Forest Service titled, "A Method for Determining Fire History in Coniferous Forests of the Mountain West." The sawyer makes two horizontal parallel cuts of equal length across half of the concaved fire-scarred area of a trunk. The two cuts should be less than two inches apart. This will be the thickness of the pizza slice. These cuts should extend all the way through the cambium (the first layer under the bark), just past the pith (at the center of the tree), and sufficiently back into the bole of the tree to ensure complete sampling of the fire scars. Next, the sawyer makes two bore (also called plunge) cuts: one along the back of the two parallel horizontal cuts and one shorter one connecting the parallel horizontal cuts where they extend just past the pith. These are vertical cuts and will form the wedge shape of the pizza slice.

The authors suggest that sawyers can obtain this sample by cutting just 10 to 15 percent of the stem. The partial cross section might not come out looking exactly like a perfect slice of pizza because of the gnarled, concave scar tissue, but the final product should be somewhat wedge-shaped. "Partial cross sections should not be made thicker or deeper than necessary," advised Arno and Sneck. "The chainsaw, a few cross sections, along with an extra quart of fuel, can be carried in an extra-large aluminum frame backpack."

When working with certain tree species, scientists actually prefer the

A polished and dated partial cross section from a fire-scarred red pine (*Pinus resinosa*) showing fire scars occurring in 1877, 1891, 1897, 1904, 1916, and 1926. PHOTO BY JAMES P. RISER II

full cross sections cut by chainsaws to plugs pulled from incremental borers. For example, the TREES Lab is working on a project with eastern red cedars that grow on small rock outcroppings in the Driftless Area. The trees have rot-resistant wood, a high tolerance for drought, and the ability to compartmentalize, which means that if one part of the tree dies, the rest of it keeps growing. The researchers are cutting cross sections of cedar stumps with chainsaws to study their history and the environmental changes that took place during their lifetimes. One cross section from an eastern red cedar growing on the talus at Devil's Lake State Park dated back to 808 AD, meaning the tree was a seedling when the Vikings started raiding the Baltic Coast and King Charlemagne had united most of Europe. For comparison, a ten-foot eastern red cedar in someone's front yard is probably between ten and twenty years old. They grow a foot or two every year.

When Larson described the partial cross section method, he noted, "It's incredibly helpful with trees that put on false rings and trees that have locally absent rings. So, with eastern red cedar, they will not put on a ring some years, or they will almost completely shut down and then start growing again. The ring sequence of cedars is really complicated and pretty tricky to analyze sometimes, so in that case, cross sections are important. There is a tremendous need for species that are challenging to work with—anything in the juniper family and some of the bristlecones. That's another role where the chainsaw makes the work easier."

14

The Soul of a Saw

The Saw as an Object of Restoration and Remembrance

A man in Manitowoc once told me that after he fixes and fires up an old chainsaw that has been sitting for decades, he feels like he has saved its soul. He smiles at the smell of stale sawdust and aged oil burning off the saw. He likes to see it come alive and rumble and vibrate on the floor. He tunes the idle until it is as perfect as it was as a younger machine in its working days. The most important gesture is shutting it off. His basement workshop becomes quiet again. There is a pause and a moment of truth; then, he yanks the pull cord. If the saw starts on the first try, it is ready for action. His work is complete. To him, this ritual is heaven on earth. Another soul has been saved.

Dan Stuckmann is as passionate as any antique aficionado, but in one way, he is different. Some people just collect, but he collects and fixes. He puts his old saws back in working order. He acquires them through hustling, networking, scavenging, and word-of-mouth. Some people bring saws to him. He explained to me in his garage workshop, "What I've found with people that I meet is they will say, 'This was my grandfather's saw. We don't have the means or interest to keep it in the family,' which I find sad, but at least they are not putting it into the scrap barrel. In those situations, if they want to see it run again, I'll always put in the effort to get it to run again, and if I can't, I can't. In most cases, if you put enough effort into it, you can get it to go."

Stuckmann grew up on a farm in Newton, Wisconsin, and he liked to use his father's Homelite XL-850. He said he never did too much tree felling or firewood cutting, but he enjoyed using chainsaws. Years later, he bought a home in Manitowoc and got a small saw for work around the house. One day, he realized he needed a bigger saw. He didn't want to spend the money on a new one, so he found a used Homelite C-51. He bought it, along with a "parts saw," for just fifteen dollars. All he had to do was clean the shellac off the points and it fired up. A point, otherwise called a contact breaker, refers to an electrical switch in the ignition system of a combustible engine that is operated by an engine-driven cam. The points open and close and are, therefore, subject to wear and tear.

"From there it kind of went crazy," Stuckmann continued. "I have an affection for two-strokes that I never had before. The next saw I got was my dad's XL-850. It was no longer wanted and still on the farm, and I resurrected that one. It exploded from there. I started picking them up here and there. The saws I got earlier were from the seventies. They aren't really collectible, to be honest, but then I started going for the big horsepower saws from back in the day."

One saw he acquired was like the mystery of Chris McCandless, the young adventurer whose body was found in 1992 in a junked school bus in the Alaskan bush under strange circumstances—a story told in Jon Krakauer's *Into the Wild*. Stuckmann found a Pioneer saw in a school bus at an abandoned logging camp near Kaukauna. The saw was drained of fuel and the chain was sharp. There was no explanation of why the owner left a well-cared-for saw behind. From that site, Stuckmann also brought home an aluminum logger's helmet, the kind used before hard plastic helmets became commonplace.

Stuckmann wanted to share his newfound work and passion, so he concocted a pallet racking system that would display his restored saws, and he began taking it to antique machine shows, thresherees, and other festivals featuring historic equipment and technologies. He quickly learned that hauling the saws and assembling the pallet racks was a lot of unnecessary work. With some help, he built a display wagon that slides into a fourteen-foot enclosed trailer for transport. The wagon's shelves are carpeted, and the saws are held in place with pipe hanger straps. Adorning the wagon are stickers and buttons from all of the shows he's attended. The largest one (and the farthest from his home) was the annual National Threshers Association show in Wauseon, Ohio.

In addition to the chainsaws and fact sheets, Dan Stuckmann's saw wagon is adorned with stickers from the various events where he's displayed it.

The wagon holds thirty-five chainsaws produced by nineteen different brands. He walked over to the wagon and pointed to some saws on the top row. "The first three years of Homelite are here," he said. The Homelite 20 MCS was the first gasoline production saw made by the company. Next to that was the Homelite 26 LCS, the second saw of that line, and the Homelite 5-30. All of these saws were discontinued in the early 1950s. "What really got me about these," he went on, "is that the carb[uretor] turns so you can buck *and* fell the tree."

He showed me a Lombard 44, a saw introduced in 1957, and explained, "Lombard took it a step [further], and they put the carb at a 45 [degree angle]. That worked but it was weird they chose that approach." This innovation—a carburetor that continues working even when turned upside down or sideways—was one of the milestones in chainsaw design. The first chainsaws were designed to cut only in one position; some could be held horizontally for felling and others could be held vertically for bucking. When a saw was tipped the wrong way, its engine would not get gas. Many companies in the fifties were on a quest to create a saw that could be held in both positions, and Lombard was the winner.

"The C-series of Homelite have a convertible drive," Stuckmann explained. "It was an optional attachment back then. It was a gear drive assembly that you could put on it, and there was a planetary system in here to gear it down. . . . All of the original saws back in the day didn't have any real rpms. They relied on a lot of torque to run big bars. Then there was a switchover. The old-timers said the direct-drive saws were no good, and they had to have a gear-drive saw. They were in the midst of that change, and Homelite came out with this [convertible drive] and it was the best of both worlds. [But] Homelite has poor records with their manufacture dates. Their serial numbers didn't mean anything. They were just numbers." For this reason, it's often impossible for a Homelite owner to know when their machine was produced.

Next, Stuckmann moved over to a smaller saw, which was the result of another industry-wide quest: the race to produce a truly lightweight saw. Homelite pulled it off in 1964. "Homelite came out with the XL-12," he told me. "Twelve, meaning twelve pounds. That was the powerhead only, and that was the first saw to be made out of lightweight aluminum and magnesium and what-not. They sold millions of those things. They were at the top of their game. It was like the old adage that nobody is going to touch us. A lot of people copied it. Back then, a twelve-pound saw was unheard of. After that, Stihl started knocking on their door, along with some other guys with other innovations, and they slowly fell down." The small XL 12 stood in stark contrast to another Homelite in Stuckmann's collection, the 7-29. In 1956, it was billed as the "most powerful one-man saw you can own." It had a seven-horsepower engine and could run a six-foot bar.

"Back in the day, in the sixties, McCulloch and Homelite were big dogs back then. McCulloch had a bunch of funky ideas like this thing," he said, walking over to a McCulloch D-30 that was introduced in 1958. "This is a center-pull start that has a Bendix that shoots out like a car starter that engages the flywheel. It's just a strange design to see it on the saw."

Stuckmann kept one saw separate from all the others in a large wooden chest on his garage floor—the only saw with its own special case. It was a

Homelite Super XL925, considered to be the best 80 cc saw the company ever made. A spare thirty-two-inch bar was mounted on the inside of the lid, along with spare chains and a measuring stick. The chest was cleverly compartmentalized to hold a gas can, oil bottle, axe, and some wedges, and it was padded with carpeting and leather. Bob Stertz of Omro bought the saw in 1980 and built the chest for it. "He grew up on a farm, and they cut a lot of firewood," Stuckmann recalled. "They protected their investment. One guy can hardly lift [that chest] because it's so long. With the saw and the gas cans, it takes two guys to lift it. [Stertz] showed me how he kept the chest on his tractor. He didn't want to part with it. He said he was too old and couldn't get it to run. There was too much compression. I got it with the promise I would bring it to shows and never to sell it."

But that's when the trouble started. Despite restoring numerous chainsaws over the years, Stuckmann could not get Stertz's saw running. He sent it to Leon's Chainsaw Parts & Repair in Oregon. The staff there determined the culprit was a decayed cylinder base gasket, which the repair shop replaced. Everything else on the saw was in great shape. The saw was shipped back to Stuckmann. He waited for thirty days. The saw never arrived. The UPS tracking number was useless. The people at the saw repair shop were befuddled, as they had shipped the saw properly. The chainsaw surfaced near Chicago when a man received an extra chainsaw from Leon's saw shop. It seemed two chainsaw shipments had been scanned as one. The saw eventually made its way back to Stuckmann. Later at an old-time farm machinery show, Stertz crept up behind Stuckmann, slapped him on the back, and asked if he had an XL925. "I sure do," said Stuckmann, not recognizing him at first. When he did remember Stertz, they both had a laugh, and there was the old Homelite—at the show with a promise kept.

As we walked around his wagon, Stuckmann observed that some of his antique chainsaws were made by companies known for other products. For instance, he showed me a Sears, Roebuck 917-60044, a saw with a four-horsepower engine. "They were just trying to be a me-too type of thing," he said of the Sears, Roebuck brand. "With Sears, a lot of their stuff was made by David Bradley. They made walk-behind power equipment and lawnmowers and chainsaws. They put their name on all kinds of products, like Kenmore."

Another one of his me-too saws was the 1958 Remington Arms

Logmaster SL5 sitting at the end of the wagon. Remington—best known for making shotguns, rifles, and ammunition—also made sewing machines and typewriters in its early years. The company bought its way into the chainsaw market by purchasing the Mall Tool Company in 1956. Arthur W. Mall founded his enterprise in Milwaukee and later moved it to Chicago and Park Forest, Illinois. "Back then it was a different time," Stuckmann observed. "Look at snowmobiles. Everybody was trying to get into that and get a piece of that. Americans were spending money like crazy. They thought, 'If we can get in on this, we'll try to be the next best thing.'"

His Remington Logmaster came with some history. According to a fact sheet about the saw compiled by Stuckmann, Ernest Wilkinsen of Cheboygan, Michigan, purchased the saw new in 1958. His thirteen-year-old son, Gerry, cut himself with the saw and ended up with sixty stitches in his left knee. Later, the chainsaw cut Gerry's younger brother David in the forearm, and he got forty stitches. Back then, saw manufacturers were slowly adding the safety features we take for granted today, and operators were relatively clueless when it came to saw safety. I doubt Gerry or David ever used any protective equipment or had any training in using their machine. Today, Remington chainsaws are made in China. By trading on the Remington name, these saws are often sold with the consumer perception that they are American-made.

Stuckmann likes to preserve the history of his saws. His Bolens 3900-1, for example, is stored with a handwritten note from the previous owner of the saw, Jim Hauser, as well as two black-and-white photos of the saw being used by Hauser's father. Hauser mentioned in his note that the photos were taken in either 1960 or 1961 at his family's home in Oak Creek, Michigan. Many of the original owners of Stuckmann's saws had inscribed their initials on the machines or their cases, along with important information such as the size of the spark plug gap and the gas-mix ratio. "A chainsaw back in the day was pretty expensive," Stuckmann explained, "and the guy who owned it took care of it."

I recognized one of Stuckmann's prized saws—a Stihl Contra/Lightning 1106—as the saw that blew the doors off the chainsaw industry in 1959. When this gearless one-man saw hit the market, it was lighter and faster than all models that had come before. It had a chipper-tooth chain and a diaphragm carburetor, which allowed the saw to be used in felling and bucking positions. "These are very sought after by Stihl collectors," Stuckmann added. "This is the saw that got Stihl back into the United States market after the war." Andreas Stihl, of course, was knocked out of the North American market during World War II when competing manufacturers were blatantly copying his design concepts. The Lightning 1106 (as it was called on the American market) was the comeback that put Stihl back in the ring and contributed to the company's domination of the chainsaw market into the future.

Of course, Stuckmann has fond feelings for some saws in his possession precisely because they were made not just domestically but locally. One was a 1946 Disston KB7-AY with a Mercury engine. The two-man saw weighed 105 pounds and housed an eleven-horsepower twin-cylinder engine. Stuckmann made a two-wheel platform on the handle end so he could move the machine around like a dolly. "That thing's a piece of iron," he stated. Carl Kiekhaefer wanted to get into military contracts. He teamed up with Disston to do the bar and chain work, while Kiekhaefer made the engine at his factory in Cedarburg. "There are a handful of saws made within eighty miles of here," Stuckmann said with pride. "Bolens is another one—these were made in Port Washington. This Wright reciprocating saw was made by Thomas Industries in Sheboygan. This is an actual Wright chainsaw. Everyone thinks when you see a Wright, it's a reciprocating saw, but Wright made regular chainsaws as well. They are actually rarer than the reciprocating saws. They had really weird designs. And McCulloch started in Milwaukee and then they later moved out to Los Angeles."

Stuckmann dragged out a beast of a saw with a scratcher chain. It was made by Disston and included the Mercury engine built by Kiekhaefer Aeromarine, founded by Carl Kiekhaefer in 1970 to build racecar engines. What was unusual about the chain was that it rode on top of the bar instead of in a groove. The chain straddled the bar, which was merely a flat stock of metal. It was an ingenious design that never caught on. He also showed me a Root, a saw with the engine made in Hartford, Wisconsin. "This was kind of an odd design," he admitted. "The air filter was up front by the bucking spikes. Not the smartest thing with a lot of dust there. Very few models made. This was the biggest one. It's loud. It's a beast."

In addition to the saws in the trailer, Stuckmann had saws scattered around on the floor of his garage and packed onto shelves. He pointed out a made-in-Canada Pioneer 620 that had a small, inverted glass sediment bowl used as a screened filtration system. Old tractors used to have these, but Stuckmann found it odd that such a fragile part would be mounted on a chainsaw. Leaning nearby in a corner was a long cardboard box holding a single-string weed trimmer called a Whizz Witch. It was made by Hoffco, and the instruction manual claimed the attachment would fit most lightweight chainsaws. Stuckmann said he had not tested it.

"Here's an[other] odd duck," he went on, showing me a huge bow-bar saw. But unlike other bow-bars, where the chain revolves around a protective rail, the chain on this saw traveled through a bent metal pipe. "This is one of my 'wow' saws when it's at a show. This is a Mall, out of Chicago. Mall was really ahead of their time. They made lots of tools that fit this powerhead. Everything from water pumps [to] generators [to] vibratory packers [or plate compactors]."

As we dug through his jumbled stack of lost treasures, he also pointed out an Ardis Quick Change Adapter, an ice auger attachment made in Cumberland, Wisconsin. Then he pointed out another ice auger chainsaw attachment made by Marvin Feldmann in Plymouth, Wisconsin. Feldmann Engineering & Manufacturing Company is known today as the maker of Jiffy Ice Drills.

Stuckmann led me from his garage through his home—where I spotted *Chainsaws: A History* by David Lee prominently displayed on a book easel—to his workshop, the place where the souls of old chainsaws were saved. Shelving racks full of saws decorated the space around his long workbench. Of all the saws in the room, I noticed that just one, on a high shelf, was mysteriously covered with a drop cloth. Stuckmann pulled it down and hefted it like he was showing some kind of prize at a county fair. It was a Homelite Super 650 with a six-star Homelite bar. He said it was a Tournament of Kings bar. It was in such good condition, it looked like Stuckmann had purchased it brand new just a few days prior. "One of my pride and joys is this thing," he told me. "It was a demonstrator saw. I didn't restore it. It's one of these saws that are in beautiful shape. I found this [when I was] looking for some old bars around Cedarburg. It was laying there covered up and I fell in love with it."

The Tournament of Kings was a big deal in its time. The chainsaw competition lasted from 1978 to 1985 and attracted sawyers from around the

CHAIN SAW AGE

WORLD WIDE SINCE 1952

OCTOBER 1981
IN THIS ISSUE
Prison Inmates Build Hot Saw
CSA Launches 'Annex' Section
Duralube Ups Marketing Effort
Shoptalk Lists Performance Mods

HOMELITE TOURNAMENT OF KINGS™
FINALISTS
1981 World Chain Saw Championship, Charlotte, N.C.

PLACE	NAME	POINTS	PLACE	NAME	POINTS
1	Ron Hartill	46	6	Ron Johnson	36
2	Roy Booth Jr.	38	7	Roy Booth Sr.	29
3	Dave Geer	37	8	Sven Johnson	28
4	Jack Jaffray	37	9	Tom Bildeaux	26
5	Larry Bildeaux	36	10	Royal Burnham	17

Like motorsports events, chainsaw contests were created to test both operator and machine. The Tournament of Kings was the chainsaw equivalent of the Super Bowl, and it garnered heavy press coverage at a time when chainsaws were being rapidly adopted by the public. COURTESY OF *CHAIN SAW AGE* MAGAZINE, HATTON-BROWN PUBLISHERS world. It was held at Carowinds Theme Park in Charlotte, North Carolina, and pulled in thousands of spectators each year. In its heyday, it was covered by NBC, CBS, *Sports Illustrated*, *The New York Times*, the Associated Press, United Press International, *Outdoor Life*, *Popular Mechanics*, *Popular Science*, and just about any other outdoor- or engineering-focused media outlet you can name. Sponsored by Homelite, it offered the biggest payout of any chainsaw event. In 1979, the prizes were $2,500 for first place, $1,000 for second, and $500 for third. Just two years later, in 1981, the first-place winner took home $6,000.

Unlike other lumberjack competitions, which include events featuring axes and crosscut saws, the Tournament of Kings tested only chainsaw skills. There were six events:

1. Bore cutting. Contestants make a bore cut through a log to puncture a balloon on the other side.
2. Felling. Contestants fell a twenty-foot spar to get it to land on a target.
3. Steeplechase. Contestants sprint through a cutting course, making fifteen various types of cuts. Winners usually accomplish this in under two minutes.
4. Precision cutting. Contestants cut a twenty-inch log in half from top to bottom.
5. Disk cutting. Contestants cut ten cookies off of a log, with the goal of having the thinnest possible stack.
6. Speed cutting. Contestants make three cuts on a log (down, up, and back down) as fast as possible, much like the competition at the Union Thresheree.

In the disk-cutting event, sawyers would cut the wafers so thin that the bar could be seen through the translucent wood fibers. In the 1979 tournament, the winning stack of ten disks measured just two inches in height. One of the most dramatic storylines in the history of the competition was a father-son showdown in 1980. Two-time defending champion Roy Booth Sr. competed against his nineteen-year-old son, Roy Booth Jr., and the young man defeated his father.

Stuckmann stowed away his baby and tucked the drop cloth back over it. Then he began showing me some of the most unusual chainsaws I had ever seen. The first, in fact, did not even look like a saw. It did not have a bar but instead had jumper cables hanging off the powerhead. "This was made by Lear Jet for jump-starting planes or cars or trucks," he explained. "It was twelve to twenty-four volt. Basically, it's just an alternator on it with a McCulloch 10-10 powerhead. It works fantastic. For the hell of it, I ran my battery down and tested it. Lear was thinking of a bush pilot that lands out in the middle of Alaska somewhere and has no power. If you had one of these in your plane, you jump-start it and get going again."

Below the Lear Charger was another odd-looking saw on a bottom shelf. It was a green Luther with a bar on the opposite side, making it a left-handed saw. "I don't know if it was meant to be like that," Stuckmann admitted. "I don't know if you would technically call it a left-handed saw. Typically, the bar is on the other side. It's just bass ackwards. The recoil is on the opposite side, which I thought was kind of strange."

He went over to a McCulloch saw on his workbench and pressed a button to make a sound somewhere between a chugging sewing machine and a Harley Davidson rumble. "It's an electric-start McCulloch," he said. "This was about 1966. These are NiCad batteries and they actually charged

This Lear Charger was originally designed for stranded pilots, but it would be useful to keep in a truck or have around the homestead. Though the portable charging unit was ahead of its time, the product never took off.

while you cut." He pointed to the batteries located under the rear handle. "The first start is by hand, but after that, when it's running, it starts right up. It didn't last that long. [McCullough] said it didn't add much more weight, but when everyone was trying to lose weight, this added weight. The flywheel and everything was different in it."

The last odd duck Stuckmann pulled out was a Craftsman circular saw with a chainsaw attachment. When the circular saw assembly came off, the chainsaw fittings were put in its place. Unlike circular saws that lay flat against a tabletop, this chainsaw attachment caused the saw bar to stick out perpendicular to the ground with the powerhead held vertically, as if a circular saw was being held against a wall. "I've cut with it," said Stuckmann. "It's not bad actually. It's really weird. I think it was meant for house framing and cutting timbers. I like what it says on the bar here: 'Oil Freely Every 15 Minutes.' I don't know if they dunked it in a bucket of drain oil or what. I always got a kick out of that. My dad had a saw like this on the farm but he didn't have this attachment."

After the workshop tour, Stuckmann grabbed his Homelite Super XL—with a bow-bar attachment—and we hiked through his backyard to his saw shed. Behind the shed and down a steep slope, Silver Creek flowed in lazy bends through a marshy area. Cedars partially blocked the view, so Stuckmann had recently taken one down and the log rested there on the ground. He propped up the log under two chunks of wood and fired up the bow bar. It was louder than most modern saws but less screechy. Stuckmann braced his left foot on the log and pushed the bumper stop against it. He leaned into the saw and gently pressed down on the bow bar. The saw easily cut through the wood, spraying chips past Stuckmann's knee. He cut a few cookies, and I could see how the ergonomics made it a breeze to buck small logs. With a modern chainsaw, a sawyer would buck by kneeling on the ground, stooping with a bent back, or squatting in a deep crouch. Stuckmann cut the engine and said, "They throw a mean chip. It's crazy the way they work. It's comfortable. Give it a try."

I picked up the saw and nudged the bumper stop against the cedar log. I gently leaned on the powerhead, adding just enough pressure for the teeth to bite into the wood. That was all it took, and the machine chewed through the wood. I cut a few more cookies and killed the engine. I had been skeptical of the bow bar, but it left me surprisingly impressed. It did

Dan Stuckmann cuts a wood cookie with this Homelite Super XL mounted with a bow bar. A bow bar allows a sawyer to stand relatively upright and lean slightly into the saw to guide the chain through the wood.

not have the overall versatility, felling capability, or quick handling of a modern saw, but if I had a large pile of small to medium-sized logs to buck into firewood-sized rounds, using this saw would certainly make the job easier on my lower back.

At the end of my visit, Stuckmann took me to one last saw storage area: his backyard saw shed, or as Stuckmann called it, the home of the misfit saws. On that day, he estimated that he owned around three hundred saws, and more than half of them were for spare parts. He has about fifty saws that run in top shape at any given time, but he admitted that number was always a moving target. He pointed out a two-man saw and said: "When I go to a show with one of the two-man saws, one of the things I get is that nobody wants to run the other end. Is there a chance a chain could snap and whip? Yeah, but it's so remote. I had a bunch of young guys last year, and [almost] none of them would do it, but one guy did. When we got done, I said, 'Think about it. How many guys have run a two-man chainsaw in this world and are still alive?' And then all of a sudden, his buddies wanted to do it too."

While walking back to the house, Stuckmann told me one attachment he would like to get was a rail grinder, a chainsaw contraption that railroads used to cut steel. His Holy Grail saw is a Homelite 7-70 GS, which has a stick shift with a two-speed gear reduction in it. He said they are very difficult to come by. "There was a woman looking at the wagon one time," he recalled. "She said, 'You know these things haven't changed in forty years.' I said no, they haven't. The concept is still the same and they may be faster, lighter, or quieter, but it's still the same tool."

Stuckmann told me a story of meeting three generations at a show, a grandfather, father, and son. The grandfather was looking at Stuckmann's saws and became excited at the sight of his McCulloch Super 44. "That's the one," the older man said. "Does that run? Do you mind starting it?" Stuckmann remembered, "He got really worked up. Just by luck, I had it running before, and so I gave it one pull and the thing is going. He pushes me out of the way, and he was recording it with his camera. The guy had tears in his eyes. He said, 'That was the second saw my dad bought . . . and to hear that sound again!'"

15

Cutting Hard Water

The Saw as a Window-Maker to Another World

While on assignment for *The New York Times* a few years ago, I reported on sturgeon spearing on Lake Winnebago. The article began: "The translucent ice on Lake Winnebago absorbed the harsh sunlight, which glowed from two rectangular spearing holes in a fishing shack's black painted walls." To me, those spearing holes were the most haunting part of the story. Those holes had an eerie surreal glow. Those holes were cut by chainsaws.

I sat for part of one morning interviewing a spearman. After I got the basic facts down, our conversation veered into other topics. For long moments of silence, we stared through the ice window and into the glowing water beneath us. It was mesmerizing in a primal way, like staring into a campfire. Dangling near the bottom of the lake at a depth of ten feet was a sturgeon decoy. It was a hand-carved fish with metal fins meant to attract curious sturgeon. A six-pronged spear also dangled in the water, its barbs below the surface so the spear would not splash when thrust. When a sturgeon passed through that window, the spearman would lift the spear off a nail and slam it into the fish. A rope knotted at the end of the spear was loosely coiled around a dowel rod, ready to spool out if the catch turned out to be a large sturgeon.

Some anglers go for years or decades trying and failing to catch a sturgeon. Not only are they motivated to avoid getting skunked, but they are also racing against other anglers. The season runs for just sixteen days, and it has a predetermined harvest cap. When a certain number of sturgeon are harvested and registered at a weigh-in station, that's it. The season is over. During the 2024 season, 147 females and 155 males were harvested on Lake Winnebago. Anglers abide by the regulations to ensure that what has now been gained is not lost. In the 1930s, sturgeon were so close to extinction, the US Fish and Wildlife Service considered adding them to the endangered species list despite the fact that they were not endangered at all in Wisconsin because of prudent management. With this grassroots conservation success story in mind, the government strictly regulates sturgeon spearing to avoid the possibility of their future endangerment. Sturgeon advocates fought back against the listing, and the endangered status was dropped. The best book on the subject is *People of the Sturgeon: Wisconsin's Love Affair with an Ancient Fish* by Kathleen Schmitt Kline, Ron Bruch, and Fred Binkowski.

Chainsaws are an integral part of the sturgeon-spearing season. Before chainsaws, ice was cut with specialized handsaws. But chainsaws can get the job done in a fraction of the time it takes with a muscle-powered saw. Similarly, powered ice augers can perforate a lake with many holes before an angler with a hand-powered auger can punch through just one hole. I've drilled many holes through a two-foot layer with a hand auger, and it is an exhausting experience. The chainsaw is also versatile for ice cutting. If the ice is thicker than normal, a longer bar and chain can be added to account for this variable.

The Wisconsin Department of Natural Resources (DNR) has rules about the size of ice holes. According to their ice-angling regulations, "No person may cut, use or maintain an ice hole larger than 12 inches in diameter or square for the taking of fish in any manner through the ice except spearing sturgeon on the Winnebago system during the open season. The total area of a sturgeon-spearing hole or combination of holes may not exceed a total of 48 square feet, measured at the surface of the ice. Ice holes larger than 12 inches cut, used or maintained for spearing may not be covered by a shelter, shanty or similar device, except during the period from 48 hours before and continuing through the open sturgeon-spearing season."

Most sturgeon-spearing holes are two feet by two feet, though it depends on the size of the ice shanty. It's best not to skimp on the size, since sturgeon can grow up to seven feet long and weigh over two hundred pounds. Sticking one with a spear and getting it up to the surface can be a

A chainsaw supported by a metal guide is used to drill a hole while fishing for sturgeon in February 2022. WISCONSIN DNR, PHOTO BY DAVID NEVALA

rodeo. The DNR also has rules requiring anglers to mark their holes with wood lathes to prevent people from falling or ditching a snowmobile into the drink.

When I visited Dan Stuckmann, I asked him about how saws had historically been used for cutting ice. He said there weren't, to his knowledge, any saws specifically made for cutting ice. Rather, existing saws were modified for the job. He told me vegetable oil was often used in the engines, so the ice and water would not be contaminated with petroleum-based oil. Stuckmann explained that the ice-cutting chainsaw users he knew "took the rakers way down, and they cut almost the entire tooth off so it doesn't drag when it hits the water. They just want it to cut ice and not anything else." He showed me a rare forty-eight-inch Homelite bar; the previous owner had ground down all the teeth for ice cutting. Stuckmann said it almost made him want to cry to see the chain in that condition, but he shrugged it off—it was just another tool tweaked to do a different job, adding character to its history.

That was how the old-timers modified the chain. I wanted to learn how modern saws were tricked out for cutting ice. There are just a few reasons for a person to cut blocks of ice from a lake with a chainsaw. So it makes sense that just a few saw mechanics in the world specialize in modifying saw chains for ice cutting. And they happen to be a short drive from each other in Fond du Lac, Wisconsin. I visited Bob's Engine Service on Hickory Street and Dave's Engine Service on Scott Street. Both companies are chainsaw dealers and sell and service other kinds of outdoor power equipment.

At the service counter of Bob's, Kory Pieper pulled out a sample chain used for cutting ice and explained how he modified it. First, every single raker is removed with a grinder. Then, the top plates on the teeth are filed off until the tooth is no long curved in a c-shape but merely a sharp fin. Finally, some of these teeth are heated and bent to the left or right. The teeth alternate in a pattern where one tooth is bent to the left, the next is bent to the right, and then the next remains straight. This repeats for the length of the chain.

Back in the Husqvarna section of the shop, Pieper showed me a Husqvarna 592 XP. Its forty-two-inch bar was protected with a lightweight metal scabbard with a Bob's Engine Service logo on it featuring an image of a sturgeon. Pieper explained how it works: "The big thing is keeping the water out of the saw when cutting ice. We seal the air covers and the vent holes on the ice saws. We put a screw greaser on it and drill the bar out so you can get more grease into the chain. We drill a hole in the side of the bar cover and drill two holes into the bar so the grease can get in there and lubricate the bar. We also use bio-bar oil, as well. It's better than vegetable oil. When you're running the chain, you want it to almost be white. Just keep screwing this in and keep pumping the grease on the chain. Everybody thinks water is a lubricant, but it's not on a chainsaw. It will ruin bars."

Pads were welded to the felling spikes on the front of the powerhead to prevent a water spray rooster tail from blasting the powerhead. "The biggest enemy out there is the water getting into the engine and washing all of the oil out of the saw and not getting the air filter wet," Pieper said. "With these saws, there is so much water running through them when you're cutting ice, and keeping water out of the air box is key. [Also, you need to] still let it be able to breathe without sealing it off completely. You can't rely on airflow from the flywheel because that's going to throw water in there."

The most visible modification on an ice saw is the grease gun mounted along the side plate. A chainsaw cutting into a block of ice for a carving contest is nothing to worry about; however, a chainsaw cutting through lake ice and taking in water will create mechanical problems

When the sturgeon-spearing season ends, anglers bring their ice saws into Bob's to be summerized, and that involves removing all water from the saw, greasing the clutch, and fogging the engine with a fogging oil. The shop also offers a winterizing service for prepping saws to cut ice.

Bob Hutter, the former owner, started modifying chainsaws into ice saws forty years ago, but there are very few customers who buy these modified saws for reasons other than sturgeon spearing. Once, a customer from the Upper Peninsula purchased one in order to cut open a river where an ice jam was blocking the passage of fish. Pieper added that the city of Fond du Lac uses them for clearing out ice jams on local waters. Also, Sturgeon for Tomorrow, a conservation group, buys two saws every year to be raffled off as a fundraiser.

A few blocks away at Dave's Engine Service, I met Travis Schneider, the grandson of the business's founder, Dave Schneider. "I remember Dad would bring home blowers for me when I was ten and I'd rip them apart," Travis recalled. "I didn't know how to fix them then, but eventually I learned. I basically grew up with all this." He gestured around the showroom of chainsaws, leaf blowers, lawn mowers, weed whackers, and other power tools.

On their wall display of Husqvarna saws, Dave's had a Husqvarna 592 XP tricked out as an ice saw with a forty-two-inch bar—almost exactly like the one at Bob's. The chain modification on the two saws was similar, but on the saw at Dave's, the greaser had a hand grip that resembled a bicycle hand brake for squeezing out the grease and the routing for the grease was slightly different.

I was impressed by the custom scabbards made by the staff at Dave's for the long bars of the ice saws. They are made out of PVC pipes that are heated and squeezed into a slot shape. "We have a mechanism my grandpa made," Schneider explained. "You heat it up and you slide the pipe in this case and you smoosh it. You let it sit until it's not hot anymore and you take it out and we paint it."

Both shops sell an average of around a dozen ice saws per year. Depending on the size of the powerhead, these can range from $1,700 to $2,000. In 2023–2024, Dave's sold just six. The winter was so warm and the ice was so thin that the sturgeon-spearing season was almost canceled. I noticed that both shops chose new Husqvarna saws as the ones to modify for ice cutting, and I asked Schneider about other brands. "It's hard to water-ize Stihl saws," he said. "We've tried, but water still gets to the air filter, so we strictly do Husqvarna. Some people request Stihl. Like, they will bring in an old one, and then we will do it. But we don't modify new ones."

Ice fishing is famously characterized by periods of time when the fish are not biting or the sturgeon are not passing through. Luckily, a chainsaw can also provide a way to pass the time: Chainsaw Frisbee. This game was introduced in 2017 by Fredrik Reuter, a Swedish logger who runs the forestry website Skogsforum. He showcased it at Elmia Wood, a forestry trade fair in Sweden. The goal of the game is to launch a wood cookie as far as possible with the tip of the chainsaw, and Reuter's contraption is something that can be made relatively easily in a garage with spare parts.

The saw is mounted on a tripod at about waist level and is set up to pivot up and down. Surrounding the chainsaw bar is a metal frame, like the guide rail on a clay pigeon thrower, that is angled toward the horizon where the cookie will be launched. The wood cookie—about the size of a clay pigeon, give or take a couple of inches—is placed flat on the metal frame and set right about where the tip of the chainsaw bar will make contact. The player then grabs the saw as normal, revs the engine, and pivots the bar upward to make contact with the cookie, which sends it flying. At the sport's inaugural game in 2017, Icelander Oskar Grönholm won by launching one 124 feet.

Another option, if the fish are still not active, would be to take your chainsaw for a ride. Swedish skier Erik Sunnerheim invented this stunt in 2015. While wearing ice skates, he crouched low with a running chainsaw, touched the lower tip of the bar to the ice, and hit the throttle. The saw pulled him around the lake as if he were skijoring with Canadian Eskimo dogs. He filmed it and shared it online. In many ways, this seems like a simple, brilliant, and fun idea. The chainsaw teeth dig into the ice and pull the saw forward—an extreme version of that grabby feeling of being pulled toward a log while cutting soft wood with a very sharp chain. The extreme danger of this stunt, however, is the possibility of falling forward onto the exposed bar and chain while the saw is running at full throttle. Lake ice is not like rink ice and is full of cracks, crust, and contours. One trip-up and the saw-skater could do a face-plant into the spinning saw teeth. The forward momentum would expose their chest, neck, face, abdominal area, and groin to laceration. Attempts at saw-skating are emergency room accidents waiting to happen. But it is a brilliant idea, nonetheless.

I believe several modifications to saw-skating could make it much safer. The first thing that stands out to me is that just the lower tip of the bar touches the ice. To make the action safer, a protective rail guard could be affixed to the top and underside of the bar, leaving the lower tip exposed to dig into the ice. There would still be a risk of falling forward at high speed into a running chainsaw, and there is no way to tell which way the exposed tip of the saw could go in the event of a spill. So, that needs improvement. But what if the bar were removed entirely? Consider the power gouger, the debarker, and the chainsaw winch—all tools that harness the power of the chainsaw for purposes other than making straight cuts through wood. An attachment for saw-skating would resemble a log rod that attaches to the powerhead. It would need to be long because a chain-driven spiky wheel at the end would be the propeller that digs into the ice. If the rod or bar was long enough, it would be no danger to the operator during a fall.

An attachment like this might even make this weird, dangerous, and ingenious idea a safer and viable form of ice travel that could revolutionize the world of ice fishing. Anglers face two main obstacles when attempting to catch fish through the ice: getting to a prime fishing spot and cutting a hole through the ice. A hole can be cut with a chainsaw, a spud bar, a hand auger, or a power auger. This is not too much of a problem, though thick ice certainly presents a problem depending on which tool you are using. Getting to a prime fishing spot, however, is more difficult. Hiking out to a spot can be time-consuming. Depending on the weather, this hike can be anything from an easy shuffle over clear ice to a death march through knee-deep snow. On massive bodies of water like Lake Winnebago or Devils Lake in North Dakota, anglers need motorized transport to get around these giant ice slabs unless they are content to fish near shore. Four-wheel-drive trucks, ATVs, and snowmobiles will get you farther out on the ice than walking could. But the ice might not be thick enough to support these vehicles.

This is when an angler with a chainsaw would thrive. This conceptual chainsaw-driven ice propeller device could transport an individual far distances across hard water. Though, of course, it has yet to be invented, I see this as another example of chainsaws' ability to empower the individual. To my mind, no other tool in the world offers such versatility and power for a person leading a self-reliant life.

16

Achieving Flow in the Forest

The Saw as Attention-Reclamation Device

The chainsaw has a magical power to solve any problem I might face. It is a temporary solution, but it works with 100 percent efficacy. With the first rip on the pull cord, all my concerns, worries, anxieties, and fears immediately disappear. Since the saw is a dangerous tool and demands total focus and attention, the operator cannot think about or dwell on anything else. Thoughts about worrisome financial problems, rusty relationships, unfulfilled dreams, political nightmares, and other existential crises are put on hold. For that moment in time, it is just me and the saw.

I often find myself craving this time as an addict might crave the next fix. I have wondered why, and I've landed on the following five reasons.

First, chainsaw work puts me in a state of flow. The psychologist Mihaly Csikszentmihalyi coined the term *flow* in 1970, and the concept has been studied with interest ever since. In the world of sports, it is called "being in the zone," a magical place where every pass is a tight spiral, every pitch is a strike, and every jab, hook, and uppercut connects. Creative people experience this, too, along with anyone else lucky enough to find an activity where they enjoy a state of immersion and total absorption in the work at hand. There is a feeling of energy and bliss. Time stands still.

The chainsaw facilitates entry into this state because it demands so much attention. The act of felling, limbing, and bucking a tree combines the factors that are thought to create the flow: intense concentration, a combination of awareness and action, a loss of self-consciousness, a strong sense of control over the situation, time distortion, and a sense of purpose provided by the action.

Second, I believe chainsaw work is an antidote for modern vampires. We live in an attention economy where a person's attention is thought to be a resource that can be exploited for profit or political machinations. Much of this psychological manipulation occurs online and is very difficult to avoid. I refer to those who traffic in attention-manipulation tactics as vampires. Vampires exist by sucking blood from their victims. Blood equals life, and life is a person's consciousness. A person's consciousness is defined in many ways by what they pay attention to. Therefore, those who attempt to capture another's attention are essentially stealing a part of someone's life. The only way I've found to avoid the attention-grabbing, consciousness-sucking vampires is to curate my own world that blocks it all out. Often, it's just best to unplug and step away from it all.

The chainsaw reclaims my attention after the endless barrage of advertising, information overload, purposefully biased news, and all of their ugly cousins. Using a chainsaw is like an exercise for my attention and focus. Running a saw is the ultimate way to recharge my mindfulness. It may sound strange to some, but there is an enormous amount of peacefulness and solitude while a two-stroke engine is roaring in your hands. It's as if the extreme noise creates a space that nothing else can enter. With a saw in my hands, I am untouchable to all vampires.

Third, chainsaw work is physically demanding. Running a chainsaw can be looked at as a grueling job or a hearty workout, and sometimes it is a blend of both. Although I train regularly in my garage gym, lifting weights and doing other calisthenics, I look at my time running a chainsaw as a bonus exercise for the day. It's a total-body workout. ESPN once put together a survey to rate the toughest sports in the world and identified ten categories that define athleticism: endurance, strength, power, speed, agility, flexibility, nerve, durability, hand-eye coordination, and analytic aptitude. They asked panelists to assign sports a rating in each category. Boxing scored the highest of all.

I would say chainsaw work in logging, firewood cutting, and arborist scenarios checks all of those boxes. Nerve and analytic aptitude are essential for felling trees. Flexibility is needed to get in position to cut on steep slopes, up in trees, or in other weird spots. Walking through slash and

I take a break from thinning out a pine grove to tighten a new chain.

millions of tripping hazards without stumbling requires agility. Hand-eye coordination is needed to make accurate, tight cuts when limbing. Durability is essential because so many things in the forest can break you. The need for endurance, strength, power, and speed is relative to how far one wants to push it.

Fourth, the work gets me outside in the woods. During the twelve years that I lived in South Korea, I noticed a cultural trend that was brought over from Japan. It was called forest bathing, or *shinrinyoku*. News reports popped up about this, and soon there were forest-bathing excursions in South Korea. The idea is to absorb the forest atmosphere by spending time in a forest. That's it. Some practitioners mix in deep breathing, meditation rituals, awe walks, and mindfulness exercises to immerse themselves in the sights and sounds of nature.

If forest bathing is quietly taking in the forest using your senses, I forest bathe all the time—with a weapon. I call it hunting. Nowadays, there are forest-bathing schools, and one can become a certified forest-bathing therapist. Sadly, the concept has been turned into a capitalist venture, and some crafty entrepreneurs are charging a premium to take people into the woods to do nothing. It is not surprising that it originated in Japan, a highly urbanized nation that has deep struggles with its connection to the natural world.

As much as I like to ridicule those grifters, I do think the forest bathers are onto something. There are countless studies that mark the physical and mental health benefits of spending time outdoors. Also at work is the idea of phytoncides, which are organic compounds released from trees and plants that are thought to have healing properties for humans. I wouldn't be surprised, since when I leave the woods after hours of work, I always feel better. The chainsaw is one reason to go into the woods in the first place.

Last, and perhaps most important, the chainsaw is a conduit for meaningful work. As I hike through the woods, I notice the areas where I've cleared invasive brush and where some native tree seedlings are making a comeback. This gives me an immense amount of satisfaction. At one spot, where the buckthorn was so thick that it blotted out the sun, I sliced it down with my saw and a sea of ferns emerged the next spring. The firewood that I harvest warms my home, and it is rewarding to create that warmth through my own initiative and energy. Watching my woodpile decrease and increase in size throughout the seasons has become a reassuring rhythm in my life. All of this meaning is brought about by a chainsaw—a talismanic tool that has changed, and will continue to change, the world as we know it.

Bar and Chain Notebook

The chainsaw is a powerful tool. Though it is not a weapon, it should be treated with as much care as soldiers treat their rifles or warriors treat their swords. I urge owners to develop a ritual of regular maintenance. Another suggestion: keep a personal policy of never lending your saw to anyone—they will never be as kind to it as you.

Guidelines for Saw Care and Maintenance

Putting On an Edge

A sharp saw is a safer saw—and a more productive saw. Thus, hand-sharpening a chain in the field is one of the most important self-reliant skills a sawyer can have. I've found that there is often an air of mystery around chainsaw sharpening. Owners of independent sharpening services know this. Many of us have seen these one-man shops in rural areas with signs in the front yard advertising the sharpening of knives, lawnmower blades, and anything else with an edge. Saw chains make up a large portion of their revenue stream. Keeping an edge sharp on a tool or a weapon is an ancient skill. As a matter of simple self-reliance, I believe this skill should never be outsourced. It is a skill that should be taught to children at the same time they are given their first pocket-knife, with the understanding that the fundamentals of sharpening a knife can be applied to sharpening other metallic edges. Many people are insecure and ignorant about this subject, and that is why one of my neighbors has a sign in his front yard and has developed a nice side business for himself.

A sharp chain will bite into the wood—sometimes so aggressively the saw seems to jump out of your hands. All you have to do is hold on and guide the bar through the cut. With a dull chain, you will use more energy pressing the chain into the wood, and this becomes tiresome. A sharp chain throws wood chips; a dull chain spits sawdust. Using a dull chain is a miserable waste of time. It is unsafe, physically exhausting, and hard on a saw. Also, there is greater chance of kickback, the operator is more likely to get tired and frustrated (leading to bad decisions), and it burns more fuel.

The simplest and most elegant way to sharpen a chainsaw is freehand with a file. Once you master this skill, you can sharpen a chain anywhere: bracing the saw against a stump in the field, for instance, or on the tailgate of a pickup truck. This method of sharpening is as easy as painting by numbers or coloring within the lines. If you can follow a straight line without wavering, you can sharpen your own chains.

The first step is to purchase chains with witness marks—an ingenious mnemonic device that builds muscle memory and acts as an instant reference point. Not all chains have them—cheap chains usually do not—and many chainsaw owners have no idea they even exist. On the top plate of each tooth on the chain is a faint angled line. This is the witness mark, and that is the angle you need to follow to put the correct edge on the tooth. Your hand file must be kept parallel to the witness mark during the filing strokes.

It is also important to have files matched to your chain. There are three common sizes: 5/32 (small), 3/16 (medium), and 7/32 (large). The goal is to create a hook-shaped curve on each cutter tooth. A too-large file will flatten this crescent shape; a too-small file will not make contact with the entire curve. After much use and resharpening, the teeth will get smaller and smaller, and that is when you can use a smaller file size to match the curve of the waning teeth.

Every file needs a handle. You may buy one in a store, but I prefer to cut my own from interesting trees I've worked on. In my saw drawer, I have nubs of musclewood (*Carpinus caroliniana*), crab apple, Hardy pear, and a black cherry burl. To make a file handle, just peel the bark off a piece of handle-sized wood, drill a starter hole, stab the pointed end of your file into the hole, and push it in deep against a hard surface. After enough sharpening, the wood grain gets a *wabi-sabi* burnish from use.

If you are sharpening your chains at home, rather than in the field, put the bar in a bench vise and, if necessary, support the powerhead by placing some scrap boards or a small toolbox underneath it. Turn the chain brake off so the chain can be rotated, and make sure the chain's tension is correct: tight enough that it won't pull out of the groove and loose enough to move back and forth. Wear gloves so you can grab the chain without worry of getting nicked.

The shortest cutting tooth on a chain is called the master cutter. The idea is that all teeth must be sharpened to match the size of this tooth, and this is the tooth that must be sharpened first. Sometimes the master cutter can be so hard to identify that people use Vernier calipers. But I believe finding the master cutter is a waste of time. Just try to keep all teeth evenly sharpened as best you can. I like to smudge a random tooth with a marker. This will tell me to stop sharpening once I've made it all the way around the chain. Husqvarna has a line of chains with a gold tie strap for this purpose. Sharpening all of the teeth on a chainsaw can be either boring or meditative, depending on how you approach it. It is easy to lose track of your progress and resharpen teeth you already sharpened. The gold link and the marker trick are helpful, but they aren't necessary—just look very closely at the teeth to identify the shiny gleam of newly filed metal.

Once you've gotten comfortable hunching over the powerhead, choose one side of the chain to sharpen first. Align your file to the witness mark and push the file forward from the interior of the tooth to the outside with steady pressure and, as Steve Miller might say, a pompatus of precision. This is not a push-pull movement—the filing should always move in the same direction. The first couple of passes will feel rough as the burrs and nicks are ground away. Each pass of the file should feel smoother and smoother as the tooth gets sharper.

A sawyer who runs multiple saws with different bar lengths and a variety of chains should devise a system to keep those chains organized. Saving the original packaging helps as a reminder.

Repeat this action on all of the teeth until you come back to the first one, and then switch sides for the opposite teeth. Throughout this process, you will want to tap the tip of the file to knock off fine metal particles that gum up the file's grooves. You will notice that your file is dull when it no longer has a grinding bite and you have to add more pressure to it during each pass. Rotating the file when making a pass is not necessary, but it will help your file wear more evenly. I've had good luck with the Swiss-made Timber Savage brand of files, and their primitive logo of a Sasquatch-like goon next to a cut tree stump always makes me smile.

After your chain has been resharpened multiple times, eventually the teeth will be worn down to the witness marks. This is the sign it's time to replace the chain. Some sawyers call these "hen's teeth." When using a chain in this condition, the kerf will be narrower and there is a chance the teeth will snap off and go flying. To get one more frugal use from old chains, I save them for the nastiest work when I know with certainty that the teeth will be damaged: cutting invasive brush out of a rock wall left by early settlers, cutting along an abandoned fence line with hidden rusty barbed wire, and so on. After that, finally, I retire them.

If a chain does not have witness marks, then it's trickier to sharpen freehand. You can rely on your previous sharpening experience and muscle memory to get the job done. But if that confidence does not exist, then it's best to work with a file guide. Pferd makes a quality pocket-sized one that can be used in the field. For those with high-volume sharpening needs—such as saw shop workers or professional loggers or arborists maintaining a fleet of saws—an electric-powered bench grinder may be very useful for sharpening chains. But loggers and arborists cannot leave the jobsite and run back to the shop every time their chain gets dull. Having a sharp spare chain along as backup is useful but having the skill to sharpen a chain in the field separates the experts from the dilettantes.

After the sharpening, your next step is taking down the rakers, if necessary. Usually this is needed about every three or four times you sharpen your chain. The required tools are a flat mill file and a depth-gauge filing plate, which is a simple piece of metal that is laid atop the chain. The tiny bit of the raker that pokes out of a slot is what needs to be filed off with smooth, even pushing strokes. *Raker* is just another word for the depth gauge, which limits how deep the teeth can cut into the wood. The rakers must be slightly lower than the cutter teeth; if not, the saw won't work as efficiently. This is evident when the saw produces relatively small wood chips. If the rakers are filed too far down, the saw will produce large wood chips, but it will become bogged down more than usual.

Dressing the Bar

Chainsaw operators always have a bar bill—a running tab resulting from continuous wear on a bar. Eventually a bar will need to be replaced. With worn-out bars, chains wobble more in their grooves and sometimes they are thrown off completely. Dressing your bar, which primarily includes removing the burrs that form along its top edge, can extend its life and save money over time. To find the burrs, run a finger along both edges on each side of your bar and the tip, feeling for ridges, nicks, and bumps. These need to be removed with a mill file. Some sawyers use belt sanders or an abrasive pad on a grinder for this task, but I believe it's good to be familiar with hand filing for when you need to dress the bar in the field. The steel of some bars is harder than others. You can clamp the bar in a vise or, if in the field, brace it on the tailgate of a pickup truck or a stump. The trick is to keep the wear on both edges even—like sharpening a knife, each side of the blade should get equal attention. Take smooth, steady strokes with the mill file and keep feeling for the burr to disappear. Once it feels smooth, you are done. I recommend checking the bar every time you sharpen the chain as part of that routine. If there is a slight burr, take a few sweeps over it with the file. Like any maintenance work, dressing the bar is better done regularly and often rather than occasionally.

You should clean out the groove of your bar as gunk builds up at the bottom. Just about any thin tool will work. I use a steel dental pick. Gunk can also accumulate in the oil hole on the back end of the bar and in the grease hole for the sprocket on the front end of the bar. Clear these with any kind of toothpick or needle-type tool you have laying around. Hold the bar in a vertical position and eyeball it down the groove. On an older beat-up bar, the groove will be narrow in some spots and wide in others. This is bad. The narrow spots pinch the chain, and the wide spots cause the chain to wobble and wander, causing crooked cuts. The Silvey bar rail closer is a handy device that mounts on the bar and fixes gaps in the groove. Two roller bearings pinch the rails back to their true gauge, and the tool is pulled along the length of the bar to form a uniform gap. To fix pinched spots, you can insert a screwdriver or a bar tool into the groove and gently twist the tool to open up the gap. I have fixed widened spots by tapping the groove with a hammer and pinching it down with pliers and a vise; however, these methods are not as precise or refined as using a proper bar-rail-closing tool.

Sometimes you see a bar on a saw with the brand-name logo upside down. Sawyers do this to wear their bars down evenly, since the bottom of the bar tends to wear out more quickly. There are two oiling holes on the bar for exactly this purpose—flip the bar one way and there is an oiling hole on the bar that matches up with the oil port. Likewise when flipped the other way. My advice is: flip the bar every time you sharpen the chain.

Using a chainsaw that makes crooked cuts is a frustrating experience. It's barely noticeable when you're cutting brush or small limbs, but as soon as you make a cut into a big log, the problem becomes evident very quickly. The kerf will be noticeably angled partway through the cut, and then the bar and chain will get stuck within the kerf itself. The chain running at the bottom of the bar and the chain at the top of the bar get pinched from the crooked angle. If you notice your saw making crooked cuts, there are a few things that could be the culprit. First, your cutting teeth could be unevenly sharpened. If this is the case, go back and resharpen everything, keeping an eye out for any tooth that seems out of place. Alternately, your bar could be bent. Hold the bar up to a bright window and look down the bar rails as if you were sighting a rifle. The groove should run true, and if any daylight is obscured, you'll know there is a slight bend in the bar.

Bars (and their matched chains) are interchangeable to an extent. A store-bought 50 cc chainsaw that comes with a sixteen-inch bar can also handle an eighteen- or twenty-inch bar. However, there is a point of diminishing returns, as the powerhead can't muster the energy to properly run ever-longer bars. A saw with a short bar is lighter and thus causes a sawyer less physical strain. However, I once worked with a foreman who preferred running a long bar while cutting brush, since he didn't have to bend so far over. If you are doing that kind of work for hours on end, every inch counts. Of course, longer bars mean more teeth on the chain and that means more to sharpen. The best way to determine your ideal bar length is to walk through your woods and size up the trees that you imagine cutting in the

Eventually, all bars will wear out and become fouled up beyond all recognition or repair. On some bars, the sprocket nose can be replaced to extend its life.

future. Let the diameter of those trees guide your decision for bar length. A Biltmore stick, a ruler used by foresters, is a useful tool that can measure a tree's diameter and height and can also estimate tree volume.

Addressing Air Flow

Dressing the bar and sharpening the chain should be regular parts of your chainsaw maintenance ritual, and you might as well clean the air filters while you're at it. On most saws, the air filter covers are held in place with a latch and can be popped off with a screwdriver. Remove the filter and wipe down the surrounding area. Use an air compressor (or an aerosol blow-off duster) to blast air from the inside out of the filter. Spray the filter with a parts cleaner and blast it again from the inside out. It is important to avoid pushing dirt and grime back into the filter from the outside. Clean the inside of the cover and add a very slight film of bar oil to the inside surface. This will act like a sticky trap for dust and particles.

Another quick maintenance task that can be done every once in a while is checking the spark arrestor. This saw safety feature was put in place to prevent forest fires. It is a small mesh screen mounted in the exhaust. When a spark hits the screen, the spark is extinguished, but it also leaves behind carbon gunk. If the screen gets too grody, the saw will have trouble running at full throttle and will have a bogged-down sound and feel. To avoid this, remove the muffler cover, then pull the spark arrestor out and heat it over a fire (with a propane torch or butane lighter, or on a fiery grill). Once the mesh is superheated, you can scrape off the loosened, charred material with a bristle brush.

Picking One Wrench to Rule Them All

Inspect the exterior of your saw and find every single screw that is exposed. Now, pop off the top cover and look for more. Identify the type of screw heads on your machine and see how many match. Chances are that many will be the same and a few might be oddballs. Most likely, the oddballs will require an Allen wrench. I recommend keeping two of these wrenches in your saw kit (with one as a backup). As part of your maintenance ritual, give each screw a firm but gentle torque to make sure none has wiggled loose. Also, toss an old toothbrush into the kit for cleaning out grime that collects in the clutch side cover.

Choosing and Storing Oil and Gas

It doesn't matter if your chainsaw is gas- or electric-powered, it still needs oil for the bar and chain. Without oil, the bar and chain will overheat and can be scorched and ruined. Sawyers have colorful discussions about the types of oil they use. Most people buy name-brand oil at the local hardware store. While this tends to be the most expensive option, it is an oil specifically designed for chainsaw use. It is tacky and stringy and has a very high viscosity. Quality bar oil should look stringy as you pour it, like warm honey. In addition to a regular summer-weight oil, some brands have a winter weight, which is thinner and pours more easily in cold temperatures. Regular bar oil on a subzero Wisconsin day pours like cold molasses.

On the cheapskate end of the spectrum, some sawyers use old motor oil for their bar oil and others use gear oil, hydraulic oil, and other lubricants left lying around their shops to make a bar oil cocktail and swear by it. They're likely to say that anyone who buys the expensive store-bought stuff is a sucker. But this is the mechanical equivalent of using Chinese gutter oil to fry your spring rolls. It can be done, but it's not right. Homemade oil cocktails do not have the viscosity of normal bar oils. Also, lighter motor oils will be thrown off the bar much faster, which means the saw will empty its oil reservoir faster. That could mean running out of bar oil before the gasoline runs out. Saws are designed so the fuel runs out before the bar oil. This is to protect the bar, since a sawyer could run a saw emptied of bar oil without noticing it.

Sometimes the bar doesn't get enough oil. One way to check oil flow is to remove the clutch cover, bar, and chain and start the powerhead. Gently rev up the throttle, and oil should ooze out of the oiler hole, a small horizontal slot. If that works, then the paper test is a good way to troubleshoot oil-flow problems with a saw. When it seems like your saw is getting some oil but not enough, the paper test will tell you the truth. Get a newspaper and spread it on the ground. Start the saw and rev it with the tip of the bar pointed at the paper. Bar oils are made not to be thrown off the bar at high speeds, so there should be tiny flecks here and there. Thinner homemade concoctions will spray everywhere. If barely any oil is sprayed off from the bar, then the problem is usually the bar itself, and the groove and oil holes are most likely plugged up.

Also, biodegradable bar oil can be found on the market, and I've heard of arborists using canola or peanut oil when they are cutting near delicate and expensive landscaped plants—they believe the petroleum-based bar oil spray might not be healthy for the plants, so they use an environmentally friendly substitute. Biodegradable oil is a good choice for a day of cutting near water. The oil spray from a saw is evident, as the oil will create a slick on the water's surface. I've never liked to see that, especially if it is a trout stream. The best way to prevent errant oil from spraying onto water is to cut with the bar pointed toward the shore, which means the sawyer must get wet—but this is a simple courtesy for those who live downstream.

I've also learned to be picky about the containers that hold my fuel and oil. Most bar oil is sold in soft plastic jugs, and I've seen many jugs punctured by sharp edges after getting jumbled up with tools in the back of a work truck. They leak overnight, and you start the workday with a sticky, greasy mess. Metal gas cans are superior to plastic jugs, and cans with EPA-compliant spouts are sadly, in my experience, poorly conceived junk. It's best to throw away these spouts and replace them with the old-school flexible-style spouts. I label all cans with fuel mixed for chainsaw use as "Saw Gas," so I don't dump it into some other machine.

In the field, nobody likes to hike any distance carrying two heavy jugs of fuel and oil that they will only partially use. There is a type of can that is simply two containers merged together, one side for bar oil, one side for fuel, and with separate spouts. This is a great product. My solution for run-and-gun chainsaw work where I expect to burn just two or three tanks of fuel is to pour my bar oil into a narrow-neck Nalgene water bottle and my fuel into a metal water bottle similar to MSR camp stove bottles. Both containers are puncture-proof, hold enough to get the job done, and are lightweight for carrying into the woods.

Since I prefer to buy local whenever I can, I use Essence bar oil made in Athens, Wisconsin. This company makes summer and winter blends sold in slim-profile gallon jugs with large handles. I get it from Wood Sales & Service in Black River Falls, a saw shop that I first visited as a boy with my father. Browsing this family-owned store with new chainsaws on the shelves set me off on a lifelong fascination with a machine that would later become an essential and cherished tool in my everyday world as an adult.

Acknowledgments

Many thanks go to Will Doyle for his in-the-field tutelage in the way of chainsaws. He is not only a top foreman but also a patient teacher and a thoughtful forester. Luke Radtke is another foreman I am grateful to have spent time with while working in many right-of-ways. Many people took time to discuss with me how chainsaws fit into their world and for that I am grateful. Special thanks to Ben Buschke, Michael Demchik, Arik Duhr, T. J. Howell, Taylor Johnson, Cyrus Johnston, Evan Larson, Kory Pieper, Ryan Schlegel, Travis Schneider, and Dan Stuckmann. I also would like thank Kate Thompson, Liz Wyckoff, and the Wisconsin Historical Society Press for believing in this project.

Sources

Books

Anderson, Leon "Buckshot." *Tall Timber, Tall Tales: Honoring the Lumbering Era and the Art of Tall Tale Telling*. Hahn Printing Inc., 2011.

Berry, Wendell. "Feminism, the Body, and the Machine." In *Why I Am Not Going To Buy A Computer*. Counterpoint, 2021.

Buchanan-Smith, Peter. *Buchanan-Smith's Axe Handbook: Knowing, Buying, Using, Hanging, Restoring & Adorning*. Abrams, 2021.

Csikszentmihalyi, Mihaly. *Flow: The Psychology of Optimal Experience*. Harper & Row, 1990.

Edinger, Gary. *Will to Live: A Saga of Survival*. Stoneydale Press Publishing Co., 2010.

Gates, Robert. *What a Life!* Self-published, 2005.

Gonzales, Laurence. *Deep Survival: Who Lives, Who Dies, and Why*. Norton, 2017.

Herter, George Leonard. *Professional Guide's Manual*. North Star Guide Association, 1971.

James, William. "The Moral Equivalent of War." 1910. In *Memories and Studies*. Longmans, Green, and Co., 1911.

Jepson, Jeff. *Groundie: The Indispensable and Diverse Role of the Arboricultural Ground Worker*. Beaver Tree Publishing, 2021.

Kinloch, Francis. *Letters from Geneva and France: Written During a Residence of Between Two and Three Years in Different Parts of Those Countries and Addressed to a Lady in Virginia*. Boston: Wells and Lilly, 1819.

Knight, Frederick. *Thorn Cottage, a Poet's Home: A Memorial of Frederick Knight, Esq., of Rowley, Mass*. Boston: Crocker and Brewster, 1855.

Kilmer, Joyce. *Trees & Other Poems*. Doubleday & Company, Inc., 1914.

Lee, David. *Chainsaws: A History*. Harbour Publishing, 2020.

Leopold, Aldo. *A Sand County Almanac and Sketches Here and There*. Oxford University Press, 1949.

London, Jack. "In a Far Country." 1889. In *Jack London: Novels & Stories*. Library of America, 1982.

MacQuarrie, Gordon. "Nothing to Do for Three Weeks." In *Stories of the Old Duck Hunters and Other Drivel*. Stackpole Books, 1967.

McCarthy, Cormac. *The Road*. Knopf Doubleday Publishing Group, 2006.

Michael, David E. *Saws that Sing: A Guide to Using Crosscut Saws*. USDA Forest Service, 2004.

Rosholt, Malcolm. *Wisconsin Logging Book 1839–1939*. Rosholt House, 1980.

Seuss, Dr. [Theodore Seuss Geisel]. *The Lorax*. Random House, 1971.

Shigo, Alex L. *Pruning Trees Near Electric Utility Lines: A Field Pocket Guide for Qualified Line-Clearance Tree Workers*. USDA Forest Service, 1990.

Silversides, C. Ross. *Broadaxe to Flying Shear: The Mechanization of Forest Harvesting East of the Rockies*. National Museum of Science and Technology, 1997.

Sloane, Eric. *A Museum of Early American Tools*. Dover Publications, 2002.

Sloane, Eric. *Diary of an Early American Boy: Noah Blake 1805*. Dover Publications, 2004.

Thoreau, Henry David. *Walden*. Boston: Ticknor and Fields, 1854.

Trouet, Valerie. *Tree Story: The History of the World Written in Rings*. John's Hopkins University Press, 2020.

Weisman, Alan. *The World Without Us*. St. Martin's Press, 2007.

Periodicals

Berry, Wendell. "Taking Draft Animals Seriously." Healing Harvest Forest Foundation. March 5, 2013. www.healingharvestforestfoundation.org/forest-understories/taking-draft-animals-seriously.

Buffett, Jimmy. "Everything in the Woods Wants to Eat a Quail (Including Me)." *Esquire Sportsman*, September 5, 1992.

"Bush Shows off His Spread, Chain-Saw Skills." *Tampa Bay Times*. August 26, 2001. www.tampabay.com/archive/2001/08/26/bush-shows-off-his-spread-chain-saw-skills/.

Congreve, Celia. "The Firewood Poem." *The Times* (London), March 2, 1930.

Fallows, James. Security Theater. *The Atlantic*. www.theatlantic.com/author/james-fallows/security-theater/.

Grenier, Richard. "Perils of Passive Sex." *The Washington Times*, April 6, 1993.

Hoyt, Shannon. "Rural Mondovi Man Shares Miracle Rescue after Tree Cutting Accident." WQOW.com, January 9, 2024. www.wqow.com/news/top-stories/rural-mondovi-man-shares-miracle-rescue-after-tree-cutting-accident/article_8f9d3e7a-af48-11ee-a1f3-9b385abfaa57.html.

King, Wayne. "A Meeting of Champions and Their Chain Saws." *The New York Times*, September 23, 1979.

Koehler, S. A., T. M. Luckasevic, L. Rozin, A. Shakir, S. Ladham, B. Omalu, J. Dominick, and C. H. Wecht. "Death by Chainsaw: Fatal Kickback Injuries to the Neck." *Journal of Forensic Sciences* 49, no. 2 (March 2004): 1–6.

Petrusich, Amanda. "Going Home with Wendell Berry." *The New Yorker*, July 14, 2019.

Skippen, Mark W., John R. Kirkup, Rosie Maxton, and Stuart W. McDonald. "The Chain Saw: A Scottish Invention." *Scottish Medical Journal* 49, no. 2 (May 2004): 72–75.

"Tree Trimmer Recovering After Chainsaw Accident in Pennsylvania." Reuters, April 1, 2014.

"Sport Skills Difficulty Rankings." Page 2. ESPN.com. www.espn.com/espn/page2/sportSkills.

Sritharen, Yoginee, Matthew C. Hernandez, Martin D. Zielinski, and Johnathon M. Aho. "Weekend Woodsmen: Overview and Comparison of Injury Patterns Associated with Power Saw and Axe Utilization in the United States." *The American Journal of Emergency Medicine* 36, no. 5 (May 2018): 846–850.

Trumble, Benjamin C., Daniel K. Cummings, Kathleen A. O'Connor, Darryl J. Holman, Eric A. Smith, Hillard S. Kaplan, and Michael D. Gurven. "Age-Independent Increases in Male Salivary Testosterone During Horticultural Activity Among Tsimane Forager-Farmers." *Evolution and Human Behavior* 34, no. 5 (September 2013): 350–357.

Weisberger, Mindy. "Europe's Earliest Form of Money Found and It's a Bunch of Rings and Axes." Live Science. January 21, 2021. www.livescience.com/earliest-money-bronze-age.html.

Wells, Robert. "Tall Tales, Tough Men." *The Milwaukee Journal*, July 4, 1976.

Woodward, George Austin. "Life in a Lumber Camp." *Munsey's Magazine*, September 1894.

Wulf, William. "Great Achievements and Grand Challenges." National Academy of Engineering's *The Bridge* 30, nos. 3 & 4 (Fall/Winter 2000): 5–10. www.nae.edu/File.aspx?id=7327&v=e3a8f2e0.

Websites

Chain Saw Collector's Corner. www.acresinternet.com/cscc.nsf.

"Chain Saw Safety." US Centers for Disease Control and Prevention. February 9, 2024. www.cdc.gov/natural-disasters/safety/chain-saw-safety.html.

Investigation Summaries online archive. Occupational Safety and Health Administration. www.osha.gov/ords/imis/accidentsearch.html.

US Department of Defense Hearing Center of Excellence. Information on the decibels of chainsaws and other noise-producing machines. https://health.mil/Military-Health-Topics/Centers-of-Excellence/HCE.

Wisconsin Department of Natural Resources. Numerous factsheets and resources about forest health, the emerald ash borer, and other tree pests and diseases. https://dnr.wisconsin.gov/.

Wisconsin Logging Museum. Various informative exhibits about logging history. Eau Claire, Wisconsin. www.cvmuseum.com/explore/museum-exhibits/47/wisconsin-logging-museum-and-logging-camp/.

Interviews

Buschke, Ben. Interview with the author. Manawa, Wisconsin. January 2023.

Demchik, Michael. Interview with the author. Stevens Point, Wisconsin. January 2024.

Duhr, Arik. Interview with the author. Baraboo, Wisconsin. December 2023.

Howell, T.J. Interview with the author. Baraboo, Wisconsin. January 2024.

Johnson, Taylor. Interview with the author. Hayward, Wisconsin. February 2024.

Johnston, Cyrus. Interview with the author. Menominee, Michigan. September 2023.

Larson, Evan. Interview with the author. Platteville, Wisconsin. February 2023.

Pieper, Kory. Interview with the author. Fond du Lac, Wisconsin. February 2023.

Schlegel, Ryan. Interview with the author. Wittenburg, Wisconsin. October 2023.

Schneider, Travis. Interview with the author. Fond du Lac, Wisconsin. February 2023.

Stuckmann, Dan. Interview with the author. Manitowoc, Wisconsin. February 2023.

Reports

Arno, Stephen F., and Kathy M. Sneck. "A Method for Determining Fire History in Coniferous Forests of the Mountain West." USDA Forest Service, 1977.

Deal, Joshua A. "Site Disturbance and Social Impacts of Animal Logging and Conventional Logging in Appalachia." Department of Forestry, Virginia Polytechnic Institute and State University, Blackburg, Virginia. www.healingharvestforestfoundation.org/uploads/1/7/0/8/17089550/joshua_deal_abstract.pdf.

"National Institute for Occupational Safety and Health Recommendations to the US Department of Labor for Changes to Hazardous Orders." US Department of Health and Human Services, 2002. www.cdc.gov/niosh/docs/NIOSHRecsDOLHaz/pdfs/DOL-recomm.pdf.

US-Canada Power System Outage Task Force. "Final Report on the August 14, 2003 Blackout in the United States and Canada: Causes and Recommendations." April 2004. www3.epa.gov/region1/npdes/merrimackstation/pdfs/ar/AR-1165.pdf.

Index

Please note that page numbers in bold type indicate illustrations.

About the Author

James Card has written for *The New York Times*, *Rolling Stone*, *The Drake*, *Foreign Policy*, and many other publications. He is the only journalist in the history of *The New York Times* to have stories about ice fishing and deer hunting on the front page of the newspaper. Between stints as a magazine editor and a journalist, he worked as a line-clearance tree faller for two years. He is the author of *The Dawn Patrol Diaries*, a fly-fishing memoir. He lives in central Wisconsin and cuts firewood with a chainsaw throughout the year.